Human Circadian Rhythms

R. T. W. L. CONROY

M.Sc., Ph.D. (Manch.), L.R.C.P.I., L.R.C.S.I. & L.M.

Professor of Physiology in the Royal College of Surgeons in Ireland

and

J. N. MILLS

B.Sc., B.Ch., M.A., D.M. (Oxon.), M.D. (Cantab.), M.Sc. (Manch.)

Brackenbury Professor of Physiology in the University of Manchester

J. & A. CHURCHILL: LONDON

1970

First Published 1970

International
Standard Book Number
0 7000.1458.6

Printed in Great Britain
by T. & A. Constable Ltd., Edinburgh

Contents

Introduction

Man must have been aware of biological rhythms from his earliest existence, since acquisition of plant food depends upon knowledge of an annual, and of animal food upon similar knowledge of a daily, rhythm, while the term 'lunatic' implies an early speculation in the field of these rhythms. With the emergence of that organized knowledge which we call science, such rhythms have been increasingly studied by botanists and zoologists, by doctors and agriculturalists, and in due course a Society was formed to bring together people from varied disciplines who shared this common interest. The Society has changed its name more than once, and is now known as the Society for Biological Rhythms. It has held a series of conferences, most of whose proceedings have been published: Ronneby, 1937 (438, 439); Utrecht, 1939 (440); Hamburg, 1949 (401); Basle, 1953 (589); Stockholm, 1955 (833); Semmering, 1957 (unpublished); Siena, 1960 (185); Hamburg, 1963 (unpublished); Wiesbaden, 1967 (799).

Several other bodies have also held conferences, and published their outcome: Ross Conference on Pediatric Research, 1961 (253); the Cold Spring Harbor Symposium on quantitative Biology, 1960 (142); the New York Academy of Sciences, 1962 and 1964 (929; 318); the Feldafing summer School, 1965 (27); the Bel-Air Symposium, Geneva, 1967 (180).

A number of books: *The Physiological Clock*, E. Bunning, 1964, English Edition (131); *Rhythmic Activity in Animal Physiology and Behaviour*, L. Cloudsley-Thompson, 1961 (148); *The Physiology of Diurnal Rhythms*, J. E. Harker, 1958 (353); *Menschliche Tag-Nacht-Rhythmik und Schichtarbeit*, W. Menzel, 1962 (586); *Biological Rhythms*, A. Reinberg and J. Ghata, 1964 (749); *Biological Clocks in Medicine and Psychiatry*, C. P. Richter, 1965 (761) and *Biological Rhythm Research*, A. Sollberger, 1965 (830), have been concerned with wider or more restricted aspects of rhythms. A few reviews of biological rhythms (26, 328, 354, 466, 914) have also appeared and one of us has reviewed circadian rhythms in man (608).

This subject is at last finding a place in systematic textbooks of physiology (612) and so, by precedent, it may in due course enter the examination syllabus.

Rhythms are widely distributed through the plant and animal kingdom; they cover a great range of frequencies, over at least twelve orders of magnitude in man, from the repetitive firing of single nerve fibres to the cycle which we each experience once only in our journey from conception to death, but which is constantly repeated in the species. There cannot be much in common between all these rhythms, and they differ both in mechanism and in their functional value for the organism. Some, such as the rhythmic beating of the heart and the movements of respiration, are critically related to the mode of action of the organ, whilst some annual rhythms may be no more than adjustment to the variation imposed by the seasons.

In this book we confine ourselves to those human rhythms with a frequency of around once a day. These have been extensively observed, and their usefulness is often, though not always, apparent; most of those who study them have graduated from the stage of amassing observations, and now attempt to explain and interpret rhythmic phenomena in terms of known physiological influences.

It is hoped that this book will be of value alike to those interested in rhythms, and to those whose interest is rather in those separate fields of study, the kidney, the endocrines, and so on, which provide some of the chapter headings. The study of rhythms can, like all fundamental scientific pursuits, bear unexpected fruit and already it is clear that human rhythms have many practical implications. When a normal value, be it body temperature or concentration of phosphate in the plasma, is known to oscillate regularly, this knowledge can aid in accuracy of diagnosis. The recognition that there are regular fluctuations in the likelihood of birth or death, in the sensitivity to a drug and even in its lethality, can also be put to good use. Again, some diseases manifest themselves by disturbances of customary rhythms, such as the nycturia of heart failure, or the isosthenuria of renal failure, and many more complex disturbances are described by Menzel (584, 585, 587).

The realization that some of these rhythms are inherent in the individual, rather than simple consequences of the alternation of night and day, helps us to understand some of the difficulties and impaired efficiency of those who have just travelled into a widely different time zone. The importance of such rhythms in those who work by night, on changing shifts in industry, or irregularly as in transport and medical services and in the armed forces, is becoming increasingly apparent.

With this great variety of readers in mind, we have arranged our chapters in different ways: some deal with one or another system of the body, and others with different forms of abnormal habits. We hope that in this way each reader can more readily follow his own interests, even though we could only achieve this at the expense of occasional repetition. At the same time, those who have been concerned only with limited manifestations of rhythmicity may welcome the opportunity to widen their horizons.

We have tried to provide a fairly comprehensive account of the subject but with the emphasis on what seems to us of importance. Likewise in the bibliography we have tried to include all the more important papers, much of it early work which is freely cited by later authors and often deserves critical perusal, but we have not attempted to cover the whole literature encyclopaedically. The persevering reader who wishes a more extensive bibliography could easily follow up the references in some of the works which we cite.

DEFINITIONS

We have kept technical terms to a minimum but those which we use need explanation and definition.

Any regularly oscillating process may be described as *rhythmic* or *periodic*, the corresponding abstract nouns being *rhythm* or *periodicity*. If the quantity studied is a continuous variable, these oscillations may ideally conform to the shape of a sine curve (Fig. 0.1(i)), which is the conventional shape for mathematical treatment; observed rhythms often depart from this form in general shape (Fig. 0.1(ii)) or in asymmetry (Fig. 0.1(iii)), which usually results in the peak and trough not being spaced equidistantly. Some rhythms consist of alternation between two discontinuous states, such as being awake or asleep, so that conformity to a sine curve can only be very approximate.

Further terms applied to such oscillations, and indicated in Fig. 0.1(i) are a *cycle*, which is the shortest part of a rhythm which repeats itself indefinitely; this can be measured from crest to crest, from trough to trough, or indeed from any convenient point in one cycle to the corresponding point in the next. The time occupied by a cycle is the *period*; this book is concerned with rhythms whose period is around 24 hours. The *amplitude* is, in strict mathematical usage, the distance from the mean value to either extreme, the peak or the trough; this definition is, however, unsatisfactory for a rhythm which is not strictly sinusoidal (Fig. 0.1(ii) and (iii)) so we are adopting the more convenient definition of amplitude as the distance from trough to peak. Care should be taken in reading the literature to avoid confusion due to this ambiguity. The *acrophase* is the time of the peak on the best-fitting sine curve; when such a curve has not been fitted, the non-technical word *peak* is used.

A rhythm may be wholly dependent upon some external periodicity, whether of environment or habit, when it is described as *exogenous*. Simple examples might be a flower which opened by day and closed in darkness, an animal whose oxygen consumption or pulse rate increased with activity and diminished with rest, or the peak in urobilin excreted after each main meal (457). If, however, the oscillations continue in the absence of such external influences, they may be supposed to originate within the animal, and are referred to as *endogenous*. When external periodicity is suspended, an endogenous rhythm may continue independently, when it is termed '*free-running*'. Some authors confine the term '*rhythm*' to an endogenous periodicity. This is an inconvenient

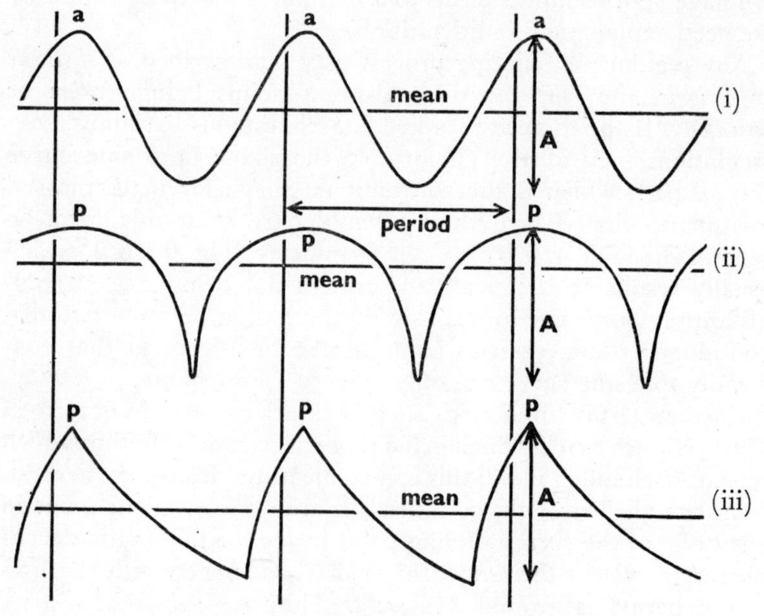

Fig. 0.1. Different forms of rhythm:
(i). sinusoidal
(ii). symmetrical but non-sinusoidal
(iii). asymmetrical
A. amplitude
a. acrophase
p. peak
The vertical lines divide the traces up into cycles.

usage since the evidence of endogeneity may be slight, suggestive, strong, or compelling; we therefore use the term 'rhythm' loosely and add a qualifying adjective when we wish to make a precise statement.

There are numerous periodic influences in our usual environment, and if one of these is dominant in determining a rhythm it is often referred to as a *Zeitgeber*, a term introduced by Aschoff (22). This can operate by a direct influence; if a plant opened only in daylight and was unaffected by temperature, humidity or other influences, daylight would be the Zeitgeber for its rhythms. In other instances the Zeitgeber may be a much more subtle and trivial influence; animals screened from most environmental periodicities but whose rhythmic behaviour was determined by the regular daily visit of an attendant to clean the cages or bring food provide such an example (116).

The *phase* of a periodicity refers to any particular point in the cycle, perhaps the maximum, or the minimum, or the point where the value rises past the mean. A *phase-shift* means that the periodicity, while remaining the same in form, is somewhat advanced or retarded, as in Fig. 9.3. The largest possible phase-shift is by half a cycle, when advancement or retardation would have the same result and, if the rhythm is symmetrical, the former maximum corresponds to the new minimum. This is often described as a reversal of phase. A phase-shift in habit and environment occurs whenever one travels east or west into a new time-zone.

The term *diurnal* has often been used in the past, and is still sometimes used, for a rhythm whose period is one day. It is, however, an ambiguous term since it is also used to distinguish day from night, and it is in this sense that we use it; the wider use can lead to such absurdities as contrasting the diurnal rhythms of a diurnal animal like a finch with those of a nocturnal animal like a rat. The term 'diurnal rhythm' has in fact been applied to fluctuations confined to the working day (274). Halberg (321, 338) has therefore introduced the term *circadian* (Latin circa dies) to indicate a period of around 24 hours. Another suggestion, diel, has the virtue of brevity, but an advantage of the term circadian is that free-running rhythms often have a period slightly different from 24 hours, which are covered by the approximating prefix circa-. In Halberg's definition (321) it covers periods from 20 to 28 hours. Some authors such as Aschoff (24) have used the term in the more restricted sense of a free-running rhythm whose period departs somewhat from 24 hours, and this has even led to a proposal to abandon the term (935); we prefer to retain it, in the wider sense in which Halberg introduced it and which corresponds to its etymology.

Another term, which in the substantive form nychthemeron has a respectably ancient provenance, is *nychthemeral*. This is most conveniently used in its strict etymological sense as referring to the alternation of night and day. Thus one may state that a particular rhythm is observed in a nychthemeral environment, implying that at least some of the customary alternations of day and night, such as of light and darkness, or of sleep and activity, are operative.

A free-running endogenous rhythm may die out after a few cycles, or at the other extreme it may persist for the whole of an animal's life. There is presumed to be some mechanism in the animal, perhaps in the brain, recording time and controlling such rhythms and commonly referred to metaphorically as the *clock*, and by continuation of the same metaphor one may talk about a *transmission system* and about the *hands*, the final external manifestations of the rhythm.

Endogenous and exogenous control commonly interact. Just as a clock seldom keeps perfect time and has to be periodically adjusted, so an endogenous rhythm whose period is not precisely 24 hours may be constantly adjusted by some factor in the environment which again is referred to as a Zeitgeber. After an abrupt environmental phase-shift, an exogenous rhythm will immediately suffer a similar shift of phase, whereas an endogenous one will persist for a time in the old phase and gradually become *entrained* to the new external rhythm. This same term *entrainment* is used for the adjustment of any endogenous rhythm to an external change whether in phase or in cycle length. A considerable vocabulary has developed around the consequences of such phase shifts but since it is used largely in work upon species other than man, it will not be here considered.

Chapter 1

Methods of Study

Rhythms in the Environment—Demonstration of Physiological Rhythms—Demonstration that a Rhythm is Endogenous—Unique Events—Transmission between the 'Clock' and the 'Hands'.

RHYTHMS IN THE ENVIRONMENT

Most men live in a world dominated by a single periodicity, the alternation of light and darkness. The day may be longer in summer, shorter in winter, but the alternation of light and darkness adheres closely to a period of 24 hours, and all our social life is geared to it. We wake and sleep, work and rest, eat and fast, drink and abstain, in a pattern which conforms to this alternation of light and darkness, and even night workers usually revert to the more customary habits at week-ends, and are in any case inevitably aware that their working and sleeping habits are out of phase with the general pattern of the community. Only in the Arctic, where day and night alternate only at the equinoxes, where summer is a period of continuous daylight and winter of continuous darkness, does man live for much of his life outside this regular 24-hour alternation, and even in many Arctic communities the social pattern still fits itself to a 24-hour clock. All forms of life which are in any way dependent upon sunlight, whether for vision, for warmth, or for photosynthesis, are to a varying extent constrained by this periodicity; since different organisms are in a multitude of ways interdependent, even those which do not themselves make profitable use of sunlight may be similarly constrained by the habits of their neighbours; thus many mammals are nocturnal, not because darkness is itself an advantage, but because it renders them less visible and therefore safer from predators. Likewise, predatory animals may be active by night simply because the forms of life upon which they prey are themselves nocturnal in their attempt to elude observation. There are, of course, many other geophysical rhythms with periods different from 24 hours, upon which some organisms are to varying extents dependent. The annual alternation of summer and winter in temperate latitudes dominates the reproductive behaviour of plants and hence often the habits of animals dependent upon plants for their food. The tidal rhythm is of great importance for animals living in the intertidal zone as for others which prey upon them; and some animals are dependent upon the lunar period governing the occurrence of unusually high tides. For man, however, the alternation of light and darkness is of primary importance, and there is some doubt whether he can even detect or be affected by other geophysical periodicities such as that in the Earth's magnetic field or in barometric pressure which have periods slightly different from 24 hours.

The chief consequence of technological advance, with the easy availability of artificial heat and light, has been that men have tended to go to bed later and rise later while still adhering to the 24-hour period. Apart from Arctic dwellers, men have only escaped from this pervasive 24-hour period by staying underground for long spells, or on submarine cruises, but this will be the natural condition of space travellers who escape from the Earth's shadow; and those who have orbited the Earth have exposed themselves to a remarkably different period, a day of some 1½ hours. A much more common disturbance, however, is a sudden change in phase, such as occurs whenever one flies some substantial distance around the world. This is now a common enough activity to raise practical problems, as well as providing much material for students of rhythms, material which may help to provide an eventual solution to the problems of night workers, shift workers, and workers in industries such as transport where irregular hours may be necessary.

DEMONSTRATION OF PHYSIOLOGICAL RHYTHMS

Since man's habits are so largely governed by the social and environmental rhythm, it is not surprising that very many physiological measurements show a more or less regular periodicity. Rhythmic variations in blood pressure, in pulse rate, in temperature, could result from a regular alternation of rest and exercise; rhythmic variations in renal excretion could result from the rhythmic habit of meal-times; variations in secretion by endocrine organs, such as adrenal cortex and medulla, could result from rhythmic variations in alertness and drowsiness, as could changes in respiration, and a variety of secondary changes could ensue; for example, changes in blood CO_2 tension and hence in pH could result from respiratory changes.

This then is the first step in the study of a circadian rhythm, the demonstration that it exists in a man living a normal nychthemeral existence. Even this seemingly simple demonstration needs some care. The two usual methods of study have been the transverse, wherein a large number of persons are examined over one or a few cycles, and the longitudinal wherein one or a few subjects are studied over a prolonged period. If one is attempting to demonstrate a reproducible rhythm, each method has its strength and weakness, and so each complements the other. The transverse design is less susceptible to idiosyncrasies of habit or reaction which may make the single subject unrepresentative of the mass of his fellow human beings, but it runs the danger of distortion by some particular climatic or other circumstance on the day of study.

Of major importance is the frequency of sampling during each 24 hours. Some variables, such as blood composition, pulse rate, or temperature, are measurements at a point in time or over a very short time spell; others, such as urinary excretion rates, extend over a longer time interval and so represent the mean level; whilst even the point measurements reflect physiological processes over a considerable antecedent time. Temperature results from the balance between heat production and heat loss, blood concentration of a hormone from the rate of secretion and of removal. Some workers have contented themselves with two measurements during the 24 hours; if these are point measurements it is obviously pure chance whether they correspond to the maximal and minimal times, or to the two moments in the 24 hours when the value was exactly at the mean and no rhythm would be revealed. One of us has shown (612) how a seemingly inverted temperature rhythm, a morning temperature above instead of below the bed-time value, can result from a small phase difference in the usual temperature rhythm. Similar considerations apply if an excretory rhythm is sought by dividing the 24-hour urine production into two aliquots. Such infrequent sampling may establish a rhythm, if it is appropriately timed and sufficiently regular, but can neither exclude its presence, nor in any way define its form. A published example may illustrate this.

In a study on a group of seven subjects over 11 days (47), blood was collected for estimation of plasma 17-hydroxycorticosteroids at 07.00 and 19.00 hours. Over the first 5 days, no rhythm was discernible, but during the last 5 days the mean morning value always exceeded the mean evening value. The authors comment: 'The absence of a circadian rhythm . . . does not indicate that the subjects had no such rhythm. It is more likely that the cycle length or the external timing . . . had changed, so that significant

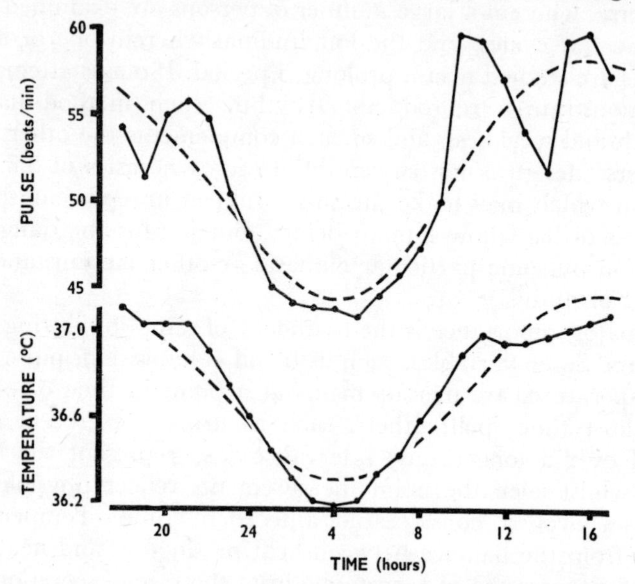

FIG. 1.1. Pulse rate and temperature over 24 hours. (data of Völker, H., 1927, 893) with best-fitting sine curves.

differences were not detectable at the habitual times of peak and trough.'

With a sufficient number of determinations in 24 hours, uniformly spaced—12 is a convenient number for computation—standard mathematical techniques may be used to fit a sine curve to the data; the variance from this curve may then be compared with the total variance, as an indication of how good is the fit to a sine curve, and an objective estimate is also obtained of the time of peak and trough, with their standard errors. This provides the simplest form of statistical assessment of the existence of a rhythm, and one which can be performed without even the aid of a mechanical calculator (924). Fig. 1.1 shows a simple example of a sine curve fitted to experimental data. Some such objectivity is necessary to escape the taunt that one is seeing rhythms where none exist, as in a study of the biological rhythms in the unicorn (152), a mythical animal whose physiological measurements can be obtained from a table of random numbers.

Where rhythms obviously fail to correspond to a single sine curve, as in Figs. 0.1 (ii) and 0.1 (iii), a better fit can be obtained by fitting several sine curves of different periods; it is doubtful if this reveals anything of physiological value, except in those instances where two influences of different frequencies are interacting. For analysis of regular but non-sinusoidal rhythms, a sine curve thus offers an ill-fitting straitjacket. When one attempts to fit a sine curve to values of a physiological rhythm, the inadequacy of fit thus arises not only from the 'noise', or irregularity in the observations, but also from the inappropriateness of a sine curve. It is questionable, however, whether it would be worth the labour to develop techniques for fitting curves of different forms such as a cycloid, which could provide a better fit to the rhythm in Fig. 0.1 (ii).

Further sophistication in statistical techniques depends upon the mathematical expertise and computer facilities available, and the appropriate manuals and original publications should be consulted (236, 272, 321, 324, 334, 335, 345, 415, 591, 831, 832). Sine curves may be fitted to long series of data, even if the records are of discontinuous states, such as sleep and wakefulness, scored as zero and unity. Sine curves of a whole spectrum of periods may be fitted, using such fine graduations as 24.8, 24.9 and 25.0 hours, in the search for the best fit of a circadian frequency which is not precisely 24 hours. Autocorrelation techniques between values 24 hours apart can establish a 24-hour rhythm whose form is non-sinusoidal. Correlation between two variables with a lag time deliberately introduced may demonstrate consistent differences in phase; this may be valuable, for instance, when comparing plasma concentration of a steroid and its urinary excretion, which commonly lags by a few hours.

Observation of a time series, reinforced by appropriate calculation, can merely establish the existence of a periodicity; it cannot indicate whether it is exogenous or endogenous, trivial or interesting. There is now a great wealth of such observations, some fragmentary, some extensive, upon rhythms in subjects living a nychthemeral existence, much of it conveniently summarized for those who can read German (586). It would indeed be hard to find any physiological measurement which has not in the hands of some assiduous investigator been shown to follow a 24-hour periodicity.

Such variations may be of considerable practical importance. Thus the physician comparing any clinical measurement with a supposed normal value will attain much higher accuracy in diagnosis if he is aware of the variations undergone in the healthy person in the course of the 24 hours. This has for long been familiar in assessing the significance of temperature; and it has more recently been appreciated that it is equally important when measuring, for example, plasma phosphate or cortisol, to note the time of day when the blood sample was collected. It is useful to know that meter readers are more liable to error (72), surgical patients to die (267), women to give birth (448), at one time of day than another, even if one is completely ignorant of the cause of such rhythmicity; and the unravelling of the chain of causation whereby some variation in either habit or environment leads to changes in a physiological variable is often a fascinating and sometimes a productive pursuit. Our knowledge of the effects of posture upon excretion of water and electrolytes arose in part from study of circadian variation, and has in turn provided clues to the control of aldosterone secretion, and may eventually lead to a clearer understanding of how sodium excretion is controlled and regulated, one of the more obscure aspects of renal physiology.

A practical difficulty in rhythm research is that it is often very laborious, since many values of any variable are needed, and it is usually much more productive to study many variables simultaneously than one alone, since their correlation, or lack of it, can shed light upon their causal relationships; and further, since observations on human subjects often involve considerable inconvenience or even discomfort, it is desirable to extract the maximum of useful information. Automated analysis, and telemetered information fed into suitable pen recorders, are helpful, and are being increasingly employed (832).

DEMONSTRATION THAT A RHYTHM IS ENDOGENOUS

If one is to understand or manipulate a rhythm, it is essential to satisfy oneself whether it is exogenous or endogenous; and this presents some difficulties owing to the pervasive character of the 24-hour periodicities, climatic and social, which surround us. It has been argued repeatedly by Brown (112-115) that seemingly endogenous rhythms are in reality the result of trifling periodic influences, such as those in the earth's magnetism and in barometric pressure. Most workers reject this hypothesis on a variety of grounds (236), among others that although remarkable sensitivity to very weak stimuli may be demonstrable in some species, as Brown claims, such sensitivity is undemonstrated and improbable in man. It is, however, desirable to establish the lines of evidence which lead to a belief in endogenous rhythms, since otherwise speculation and investigation upon the nature of the biological 'clock' is irrelevant.

The main lines of evidence have been outlined elsewhere (608). Ideally, one would like to observe the persistence of the rhythm in the absence of all circadian clues. This implies that an individual or community live in complete isolation from social contact, without a timepiece, insulated from sounds indicative of the rhythmic habits of human communities, with daylight and temperature fluctuations likewise excluded. Such, or a near approach to it, has been achieved by a succession of subjects who have spent months in solitude in deep caves, and by subjects studied by Aschoff and his colleagues more comfortably confined in a well-equipped deep underground bunker (29, 30). The same technique is easier to apply to other species, who are offered no choice in the matter, and have been maintained in such isolation from external periodicities for several generations, or have been reared from the egg under constant conditions (24). Under such conditions men usually adopt more or less rhythmic habits of eating and sleeping, in some sort of conformity with usual nychthemeral habits. It is thus not easy to decide whether they are impelled by an inherent rhythm of sleepiness and wakefulness, or of appetite and satiety, or of some other variable which determines the rest of their rhythmic behaviour. It is also conceivable that they sleep because they are tired with activity, mental or physical, and awake because they are rested and refreshed.

A less rigorous attempt at constant conditions is often attempted, by excluding those factors supposed to influence the variable in question. Renal excretion of water and most solids is known to be affected by posture, so a subject can remain recumbent, or seated and gently active, for one or several days; excretion may be affected by eating and drinking, so he can divide his intake into identical aliquots equally distributed over the 24 hours: four, six, or eight meals at intervals of 6, 4, or 3 hours, or a snack every hour, have been adopted by different workers. Sleep again can be dispensed with for 24 hours or more, and it is easy to maintain continuous illumination. While such techniques do not produce rigid evidence for endogeneity, they do at least demonstrate the independence of rhythms from the more obvious environmental influences apart from the psychological effect of the activity of other people which, judging from many observations upon night workers, can be important in man.

Since our habits of eating and sleeping make it impossible to maintain uniformity for long, another method of study is to rearrange these habits so that they do not follow the customary 24-hour cycle. One may live on a 12-hour day, so that every influence of sleep, meals or other habit is precisely reproduced at two points in each 24 hours (603, 613, 686). One may live on a day slightly longer or shorter than 24 hours, and avoid the external 24-hour influences by doing this in an isolated community either in a deep cave (475) or in the continuous daylight of an Arctic summer (513, 514, 515, 517, 826). One may completely randomize one's habits (523). The only remaining clues to the passage of 24 hours are either those meteorological tides which are probably below the threshold for human detection, or, in the Arctic summer, the nearly horizontal passage of the sun around the horizon, if it is not obscured by cloud.

Another form of artificial time-shift is to change the phase rather than the period of one's rhythmic habits. In the Arctic summer, this may be most dramatically achieved by a 12-hour phase-shift, or phase reversal (570, 807-813). It is also achieved every time one travels round the world into a different time zone; if one travels slowly on a ship or overland, this is equivalent to living on a day slightly longer than 24 hours if one travels westwards. Rapid travel by air causes an abrupt shift. This is much the most clear-cut form of phase-shift, since all the meteorological and social rhythms will undergo a similar change, and it is of great practical importance, since most people who fly long distances do so to engage in learned or business conferences, athletic competitions, or other activities in which a high level of mental or physical performance is desirable. For the same reason, opportunities for detailed and controlled physiological study are few, since the travellers are usually otherwise preoccupied either immediately before or after their journey, or more commonly both.

When in due course a physiological rhythm has become entrained to some new time schedule, whether to a change of period or of phase, one may revert to the previous nychthemeral habits, in phase with the environmental periodicity. If the entrained rhythm persists, at variance with all the external rhythms, convincing evidence is offered that the new rhythm is endogenous, that the entrainment has somehow affected the 'clock' which eventually conforms to the new time regime. It is probable that in man such entrainment to a nychthemeral existence during infancy is responsible for the development of circadian rhythms, since they are not present at birth, normally appear during the early months of life and seem never to develop in Eskimos; these people are exposed to nychthemeral alternation only for a short period around each equinox, and during the summer lead a completely non-rhythmic existence (533, 536), unlike urban communities such as that of Tromsö who, even in the continuous daylight of the Arctic summer, continue to sleep and work by the clock, so that at midnight the shops are shut and the streets deserted (475).

Another piece of evidence for the endogenous nature of rhythms comes from the behaviour of animals, or men, isolated from all overt external periodicities. Although various physiological functions remain periodic, the period usually departs somewhat from 24 hours, so that their behaviour becomes increasingly out of phase with day and night; it is to include such rhythms whose period departs slightly from 24 hours that the term 'circadian' was introduced. It has been suggested that such rhythms represent a dependence upon some geophysical rhythm other than the alternation of light and darkness; but the physiological rhythm often has a period corresponding to no known external rhythm, and has a somewhat unpredictable character in that it is slightly different in different individuals, so that subjects, human or other species, may become desynchronized from one another (725, 728). The shift is however predictable in direction according to Aschoff's rule (23), which has been verified on a number of species: in a nocturnal animal dim light shortens and bright light lengthens the period, whilst a diurnal animal behaves in the opposite way. The obvious interpretation, accepted by the great majority of workers, is that the 'clock' does not keep perfect time, and specifically tends to gain or to lose according to the prevailing level of illumination, but that in a nychthemeral or other rhythmic environment it is repeatedly adjusted by one or more Zeitgeber in the environment (see also pp. 97-98).

UNIQUE EVENTS

A special note is needed on those rhythms to whose study each individual can only contribute one point. This may be because the measurement involves killing the animal, to determine concentration of hormones in the endocrine gland which produces them, of enzymes in organs, of chemical transmitters in different regions of the brain; or it may be because the event only occurs once in the life of the individual, such as eclosion of insect pupae, birth, or death, or because the opportunity only once presents itself: mitotic rate has been measured in preputial skin removed at circumcision. A rhythm can only be expected if the population is synchronized, by either an external or a mutual influence. If we accept that free-running rhythms will drift unpredictably from precise 24-hour time, then the individuals in which they occur must become increasingly desynchronized until, with each contributing a single value, no rhythm would be discernible in the population as a whole. The failure to observe convincing rhythms, for instance in liver-glycogen concentration of chick embryos (711), is no evidence against the existence of circadian rhythms in these birds. The clear-cut appearance of a liver-glycogen rhythm after hatching may be ascribed to the potent synchronizing influence of, most notably, the alternation of light and darkness. The demonstration of free-running rhythms of periods differing randomly from 24 hours is thus hardly demonstrable in such functions unless one starts with a perfectly synchronized population and observes their behaviour over a limited number of cycles before desynchronization is complete. Synchronized rhythms in a population must therefore depend upon the continued operation of a Zeitgeber, and purists object to the application to such of the term 'circadian' (935).

We may suppose, however, that the rhythm of the unique event in the population represents in the individual a rhythmic change, a regular circadian oscillation in, for example, the likelihood of death such that its incidence clusters around a particular clock time. Likewise one supposes, when one demonstrates a rhythm in the enzyme content of an organ by killing individuals at different clock times, that the same rhythm was proceeding in each individual. If so, the rhythm could free-run in the individual isolated from external periodicity, and drift from clock time. Such a phenomenon could not be rigidly demonstrated; but it would be strongly suggested if sufficient groups of individuals were examined in a nychthemeral environment, and the shape of the peak demonstrated. If further groups were then placed under constant conditions, and sampled after a few days, a week, two weeks, and so on, it might be expected that, with increasing desynchronization, the peak would become gradually broader and flatter, until at length the variable studied was uniformly spread over the 24 hours. At this stage of complete desynchronization, it would be impossible to say whether the individuals were still rhythmic, but observation of the intermediate stage would suggest that the loss of rhythm was due to desynchronization, especially if the total range of individual values was as wide as on a nychthemeral existence.

Of the position, and nature, of the clock little can be found by human experiment, though some clues have been obtained by observing the result of lesions of the central nervous system. Observation can usually be made only of the final manifestation of its workings, the 'hands'; thus one may for example observe the vagaries of the rhythm of body temperature and try to infer therefrom the operation of the clock mechanism whereby body temperature is controlled. Intimate dissection of the cellular mechanism can only be performed upon lower animals (223, 359, see Chapter 8).

We may now summarize the different lines of evidence that any observed rhythm is endogenous:

(*a*) It persists in the absence of all rhythmic influences likely to affect it.

(*b*) It retains a roughly 24-hour period when a different period is adopted for the rhythm of habit and environment.

(*c*) It does not immediately change its phase after an abrupt shift in the phase of the rhythm of habit and environment.

(*d*) If it has become entrained to a new period or phase, as in (*b*) or (*c*), it does not immediately revert to its former period or phase when the former habits and environment are restored.

(*e*) When Zeitgeber are sufficiently excluded, as in (*a*), its period drifts somewhat away from 24 hours.

TRANSMISSION BETWEEN THE 'CLOCK' AND THE HANDS

Investigation of the transmission mechanism between the clock and the hands is beset with pitfalls, but apart from its inherent interest it is probably needed if we are ever to manipulate these rhythms for our convenience. So many functions vary circadianly that a regular phase relationship between two of these is no more than suggestive. If, however, the same temporal relationship is observed in other circumstances, then the likelihood of a causal connection increases. One may observe that the peak of plasma cortisol concentration precedes the sharp fall of eosinophil count by about 3 hours. If one further observes that an injection of cortisol is followed after a similar interval by a fall of eosinophil count, (387, 668) one begins to suspect that the adrenals are responsible for the eosinophil rhythm. A strict proof that the cortisol rhythm is the unique and sufficient cause of the eosinophil rhythm would demand that the eosinophil rhythm is absent in adrenal deficiency (349), and is not restored by giving adrenal hormones uniformly spaced over the 24 hours (446); that a dose of cortisol about equal to that normally liberated from the adrenals in the morning will induce an eosinopenia of appropriate magnitude; and that a similar dose given at any other time of day will induce a similar eosinopenia. Without this last demonstration, it could be that the sensitivity to cortisol was varying circadianly due to the rhythmic influence of some other hormonal change, so that only at the appropriate time of day would a physiological dose of the steroid exert the usual effect. The extent to which these criteria are satisfied, for different components of the rhythms, will be considered in subsequent chapters, but the evidence is seldom complete.

It is much easier to obtain negative than positive evidence in such studies. If one imposes conditions likely to disturb rhythms, such as a constant environment conducive to free-running, or a shift of phase or of cycle length, rhythms will gradually shift, and may become entrained to the new habits. If the time taken for entrainment of different rhythms is noted, it often appears that two of them become dissociated; sodium excretion may become entrained to an artificial 21-hour day whilst potassium excretion adheres to a 24-hour cycle. These two manifestations of renal rhythmicity cannot therefore have a very tight causal nexus, and in particular neither can be the sole and sufficient cause of the other.

A common method of study has therefore been to observe the rhythms after a sudden phase shift, or in subjects living on a day of abnormal length, and to note the successive entrainment of different rhythmic functions. If for instance temperature rhythm adapts rapidly, sodium excretion more slowly, and excretion of potassium and corticosteroids more slowly still, it is often suggested that temperature and sodium excretion have little to do with one another or with either of the other functions, whereas potassium excretion may be related to adrenocortical function. This supposition is not entirely justified, since many influences besides the clock can affect the measured functions. A simple example may make this clear.

Let us suppose that a morning increase in cortisol production contributes to a circadian rise both of temperature and of potassium excretion, and that the adrenals retain their previous rhythm for some time after a sudden phase reversal. By itself, this would preserve the phase of both the other rhythms under consideration. However, let us further suppose that the subject reversing his phase was in the habit of taking a swim in cold water in the evening, and a hot bath on waking, and that these two immersions alter deep temperature independently of the rhythm due to the adrenals, but have no effect upon potassium excretion. Under normal conditions his bathing habits and his adrenals would affect his temperature in the same way; if he abstained from bathing his temperature rhythm would persist and so be described as endogenous. On phase reversal, if he continued his usual bathing habits, temperature would appear to adapt immediately, and potassium excretion slowly, and the false conclusion would be drawn that their circadian rhythms were not causally connected.

Observation of a rhythm at variance with the influence of environment and habit is always more significant than that of a fully entrained rhythm. Observations of association and dissociation are in fact only pointers to promising lines of study; and it must never be forgotten that there are many direct environmental periodic influences, apart from any operation of the biological clock: we are tired at night not only through an inherent rhythm, but also because we have had a busy day.

Chapter 2

Temperature

Early Observations—Diurnal Rise and Nocturnal Fall—Individual Variations—Control of the Temperature Rhythm—Inversion—Adaptation to Phase-shifts—Effects of Isolation—Endogenous Nature.

A rhythmic variation in body temperature during the course of the day was recorded in detail by Davy in a paper to the Royal Society in 1845 (179). Davy recorded his own temperature orally every two to three hours between 07.00 and 01.00 hours and, as, he was aware of the effects on body temperature of external heat and cold and of rest and exercise (178), he stayed indoors during the period of the observations confining himself to rooms of about the same temperature and keeping his degree of physical exertion at a fairly constant level. He found that his temperature under these conditions varied from 97·6° F at 07.00 hours to 98·9° F at 16.00 hours. In another investigation which extended for eight months, he noted his temperature on rising, before going to bed and often in the afternoon as well. In addition to the classical observation that the mean daytime body temperature taken orally is 98·4° F, he noted that during the day the temperature is higher than at night and made the prescient comment that the low values seen at this time were all the more remarkable 'as the temperature of the room in which the observer sat at night was almost uniformly higher than of that which he occupied during the day'. By the frequency and timing of his observations and his efforts to maintain constant environmental conditions, Davy set an example which still holds good in the field of biological rhythm research.

During the course of the next half century a very large number of observations on daily changes in body temperature were made by many investigators (10, 35, 36, 174, 402, 422, 444, 484, 646, 683, 703, 763). Baerensprung, 1851, 1852 (35, 36), took records during the night, as well as in the daytime, and fixed the time of lowest body temperature as 04.00 hours and of maximum as being between 18.00 and 19.00 hours. Damrosch, 1853 (174), observed that the temperature rose from 07.00 hours to 10.00 hours and then fell until 13.00 hours before reaching a maximum about 17.00 hours. Ogle, 1866 (683) and Allbutt, 1872 (10) also found nocturnal temperature readings to be lower than those taken during the day.

The existence of a diurnal rise and nocturnal fall in body temperature gradually came to be accepted and its clinical value appreciated although observers differed as to the precise times at which minimum and maximum temperatures occurred. These variations were probably due to different habits of life on the part of the subjects studied and to the large individual variations in pattern which are now known to occur. The consistency of the rise and fall in temperature in a single individual was, however, very well demonstrated quite early on by Ringer and Stuart, 1877 (763). They recorded the axillary temperature every hour in a twelve-year-old boy who was in good health, and who was confined to bed during the course of the observations. Records were taken over a period of 50 hours thus enabling two complete cycles to be charted and were found to be very similar in both phase and amplitude.

An early experimental investigation into the control of the temperature rhythm was performed by Ogle, 1866 (683). He discovered that the rhythm persisted during bed rest throughout the day and in disease, whether or not accompanied by fever. He also found that the morning rise still occurred even when light was excluded. Other observers (422, 484) noted that working at night and resting during the day could disturb or possibly even invert the normal temperature curve. Mosso, 1887 (646), in a series of observations taken before and for a few days after changing from day to night work, observed that the morning rise still occurred but that there was a fall following sleep during the day and that this fall was more marked after four days, suggesting a gradual adaptation to the new schedule of activity and sleep/wakefulness.

The time needed to adapt is the probable explanation for apparent failures, such as those reported by Benedict and Snell, 1902 (59), to produce inversion on reversing the normal sleep/wakefulness and activity routine. Another early example of the persistence of a pre-existing biological rhythm, when the subject remains linked to a nychthemeral existence, is shown in Benedict's, 1904 (58), observations on a nightwatchman who was found to have an unadapted temperature rhythm. The subject had been a nightwatchman for many years but he also had a daytime job. Nightwatchmen and nurses, though frequently chosen as subjects in studies of the effects of inversion of physiological rhythms, are, in fact, particularly unsuitable for this purpose, the one tending to have periods of sleep at night and a daytime occupation, and the other to make strenuous efforts to maintain a normal social life.

During his massive studies upon sleep, Pieron and his colleagues (716, 874) also investigated inversion of the body temperature rhythm and remarked that this change was often shown by night workers. The influence of sleep on the body temperature was also remarked on by Bardswell and Chapman, 1911 (40) and by Lindhard, 1911 (524). Reviewing the results of previous investigations and his own observations, Pembrey, 1898 (703) concluded that the daily variation in body temperature was a feature of corresponding variations in the activity of the body tissues but that there appeared to be 'a certain periodicity, the result of long continued habits of life, stamped upon the processes which regulate temperature'. Woodhead and Varrier-Jones, 1916 (932), in another review of earlier work noted that sleep and light influence the temperature curve and that the rhythm persists even when subjects are recumbent in bed for 24 hours or more.

A circadian variation in heat production has been shown to occur not only as a result of a circadian variation in muscular activity and food intake, but also in fasting subjects kept strictly resting, and this fluctuation corresponds to the body temperature rhythm with lowest values in the early morning (89). Since body temperature depends on a balance between heat production and heat loss, it might be suggested that the circadian periodicity in heat production could be responsible for the body temperature rhythm. This, however, is unlikely for the amplitude of the temperature rhythm can be as great in persons who are confined to bed as in active subjects undergoing far greater metabolic variations. It is possible that the metabolic rhythm is a consequence rather than a cause of the temperature rhythm. The amplitude of the metabolic rhythm accords, in fact, with what might be predicted from the van't Hoff Arrhenius equation from a daily temperature variation of about 1° C but although it has been suggested that minima in body temperature and in metabolic rate may coincide (182), attempts to correlate temperature curves very closely with variations in energy metabolism have been unsuccessful (593).

Variations in the heat-loss mechanisms, such as cutaneous blood flow, are probably more important in regulating the body temperature rhythm and it has been shown (338, 386) that the daily variation in skin temperature is almost a mirror image of the change in rectal temperature. The 'critical skin temperature', the cutaneous temperature below which if the room temperature falls heat production must be increased linearly to maintain the body temperature, shows a rise at night. So, too, does minimal conductance, measured by dividing the basal heat production by the difference between the body core temperature and critical skin temperature, suggesting a correlation between vasomotor tonus and the body temperature rhythm (28). The amplitude of the body temperature rhythm is, however, very similar both in the humid climate of the tropics (5, 172) and in the cold of the Arctic (524) to that in temperate climates. Thus it is unlikely that the heat loss mechanisms are primarily responsible for the circadian periodicity in body temperature. It seems more likely that the rhythm exists either in the temperature regulating mechanism as a whole or at the hypothalamic level rather than in any one component of the heat production or heat losing mechanisms.

Parallelism between evening and morning temperature during the menstrual cycle has been noted (173, 476). The circadian variation in body temperature is, however, quite sufficient to obscure the difference between a woman's basal preovulatory and postovulatory temperature (691).

A negative correlation between body temperature and blood eosinophil levels has been reported as has parallelism between circadian rhythms in temperature, urinary diuresis and 17-ketosteroid excretion (594). Haus and Halberg (361), in a prolonged series of measurements on a single adult male subject, made the interesting observation that the circadian crests in 17-hydroxycorticosteroid excretion and 17-ketosteroid excretion show a consistent phase lead from the subject's oral temperature rhythm.

The existence of a characteristic pattern of body temperature and heart rate for a given individual has been emphasized by Kleitman and Ramsaroop (476). The time and duration of maximum, in particular, may vary widely from person to person but tend to be constant in any one subject (Fig. 2.1). Sex differences in the average times of maxima and minima have been suggested (580). The body temperature curve may also be related to the psychological make-up of a subject and to his mental performance during the course of the day (74, 75, 477). 'Morning'

FIG. 2.1 Oral temperatures of six male subjects taken at different hours of the day and night. Each horizontal bar represents the number of times a particular temperature value occurred. A and B are from the same subject, as are C and D. (Kleitman, N., 1963, Fig. 15.6, 470.)

types tend to have a temperature curve in which peak values are reached early in the day when subjectively they feel at their best. Conversely, evening types are likely to show a rise in temperature during the day (470) (see also Chapter 7).

Individuals may also differ in the rate of adaptation or change of their temperature to a new schedule of sleep/wakefulness or activity. In Kleitman's classical experiments on two subjects who attempted to live on a 28-hour day whilst under conditions of isolation in a cave in Kentucky, the temperature rhythm of one subject adapted quite readily whilst that of the other did not adapt at all during the 32 days that they spent in the cave

(Fig. 2.2). Similarly, as one of us has shown (610), in a group of seven subjects, subjected to a 12-hour time schedule over 48 hours, three subjects maintained their pre-existing 24-hour temperature rhythm, in one a 24-hour and 12-hour components were visible, and in the remaining three adaptation to the 12-hour schedule

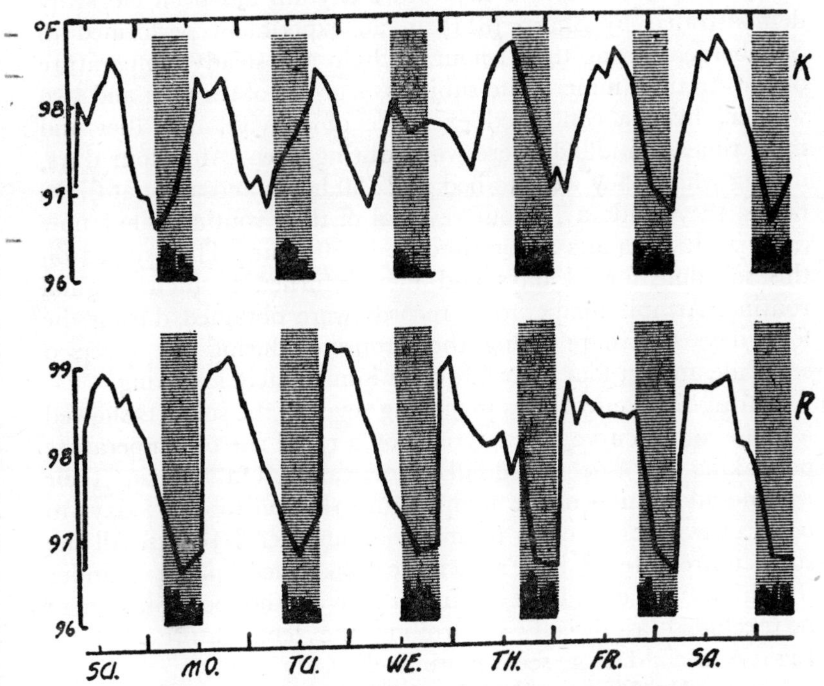

Fig. 2.2 Temperature curves of two subjects, K and R who were attempting to follow a 28-hour schedule consisting of 19 hours wakefulness and 9 hours sleep. K shows seven 24-hour curves, but R has only six curves per week with minima always in the shaded areas, which represent time spent in bed, i.e. he has adapted to the 28-hour schedule. Each weekly record has been based on the mean temperature values of the last three weeks of the experiment (Kleitman, N., 1963, Fig. 18.4, 470.)

occurred. Differences in the rate of adjustment of different individuals after flights across time zones have also been noted. The age of a subject may be important too in such changes. Sasaki (789) has noted that the temperature rhythm of a 6-year-old child entrained much more rapidly to the new environment after a flight from Japan to the U.S.A. than did the rhythm of the two adults who accompanied him. In general, adaptation of the body temperature to a phase-shift in the order of 6 hours following a rapid time zone change, such as a transatlantic journey, takes several days (133, 161, 164, 362, 363, 847). In sea voyages the body temperature appears to follow easily the slow shift in time (163, 285, 682, 687).

Reversal of the body temperature rhythm has been elegantly demonstrated by Sharp (811) in an experiment performed in Spitsbergen during the 24-hour daylight and steady temperature of the Arctic summer. The subjects, a group of six men and two women, lived a controlled existence of activity, rest, diet and meal times. Blindfolds were worn during sleep. After four days, during which they went to bed at 22.30 hours and arose at 07.00 hours, they made a 12-hour reversal of their routine. They now arose at 19.00 hours and retired at 10.30 hours. They stayed on this schedule for 14 days and then returned to their original routine. Armpit temperature records were obtained during the four days before reversing their routine, during the reversed schedule and on four alternate days immediately following.

During the control days preceding reversal the subjects showed a temperature curve characterised by a rapid rise of temperature on waking and a nocturnal fall to low values. On reversing their routine adaptation of the temperature rhythm took 3-4 days to occur. Reversion back to normal took another 3-4 days. All the subjects are stated to have shown the same general pattern. Under the circumstances of the experiment activity and food intake may be excluded as directly controlling the temperature rhythm. Light or social factors seem more likely Zeitgeber.

Burckard and Kayser (132) by altering the hours of illumination and feeding by 12 hours, were able to show a reversal of body temperature in a 27-year-old bedridden idiot. This took six days to occur. On a 6-hour dark, 6-hour light routine with feeding at 3-hourly intervals, his body temperature returned to a 24-hour rhythm.

The role of higher nervous centres in maintaining the temperature rhythm has been studied by Kleitman *et al.* (478), who found that in a subject unconscious for over a year from a brain tumour, the normal rhythm was lost. Circadian rhythms in body temperature have, however, been noted in subjects after subtotal hemispherectomy. Neither the cerebral cortex of either side nor the caudate nucleus or globus pallidus of one side appear essential for the preservation of the rhythm (337).

The development of the body temperature rhythm in childhood and its close association with the development of the sleep/wakefulness cycle has been examined in some detail by several observers and will be discussed in Chapter 8. Aged subjects have been shown to have a clearcut circadian variation in body temperature with normal phase and amplitude variations (539).

That the temperature rhythm has a free-running cycle of rather more than 24 hours is suggested by the evidence of Aschoff and his colleagues (29, 30, 32). In one experiment they confined a group of subjects for 8-19 days in a bunker and instructed them to attempt to follow regular habits in eating and sleeping in the absence of a timepiece. Eight of the nine subjects adopted a 'day' slightly longer than 24 hours with a body temperature still varying rhythmically but with a cycle also of more than 24 hours.

In another series of experiments rectal temperature records were obtained using continuous recording apparatus from subjects who were confined to the bunker, either with constant illumination and with room temperature at the subject's choice but with all known Zeitgeber excluded, or with a 12-hour light 12-hour dark cycle as a Zeitgeber. Measurements were also obtained with subjects living in a climatic chamber with constant temperature, aware of the time of day and controlling the period of lighting. Normal activity and rest were permitted in all three sets of circumstances. When Zeitgeber were not excluded both body temperature and activity rhythms were synchronised to 24 hours; the body temperature maxima occurred in the second half of the time spent in activity and the minima in the latter part of the rest period. When Zeitgeber were excluded a very different pattern was seen. The body temperature and activity cycle showed free-running circadian periodicities with a mean duration of 25.3 hours; the temperature maxima, in nine out of ten subjects studied, occurred in the first half of the activity cycle and the minima, in all ten persons, at the end of the activity period or during the first half of the rest period (29).

Fifty subjects kept in conditions of continuous illumination and each completely isolated also showed free-running circadian rhythms in body temperature, activity, sleep-wakefulness and renal function. Thirty-six remained internally synchronised throughout the experiment, fourteen showed some desynchronisation of function. Of these, nine subjects had different circadian frequencies in activity and body temperature. In two of these individuals desynchronisation began immediately after being enclosed in the bunker, in the other seven subjects after 9-23 days (30).

Similarly the body temperature of an adult female subject who remained in isolation for three months showed a free-running circadian rhythm with a period of 24·5 hours (750). Two subjects studied whilst in isolation for nine days in a constant environment also became free-running, one having a cycle length of 24·6 hours and the other 23·8 hours (790). Aschoff and his colleagues suggest that the temperature and activity cycles act as two coupled oscillators and that when Zeitgeber are excluded they begin to oscillate with spontaneous circadian frequency and a changed phase angle difference.

A prolonged series of observations on the temperature of a subject who was in isolation for a period of six months has recently been reported (153). The subject, a 25-year-old male adult, remained in a tent in a deep cave and had no means of determining time. Rectal temperature was continuously monitored for periods of up to a month over a total of three months. Records were made half-hourly so that 48 measurements were obtained daily. In all 82 cycles were measured. The subject's temperature gradually shifted so that after 160 days the shift was over 120 hours. Analysing the data by studying this shift, a mean period for the temperature rhythm of 24 hours 28 minutes was seen in the cycles for the first 2 months as against a period of 24 hours 44 minutes during the last four months.

From these experiments and observations it would appear that the circadian rhythm in body temperature is an endogenous rhythm which satisfies the various criteria for such a rhythm as described in Chapter 1.

Apart from its clinical relevance the body temperature rhythm has also been used in studies of adaptation to shift working (see Chapter 10) and may form an important link between physiological and psychological variables (see Chapter 7).

Chapter 3

Endocrine Rhythms

The Pituitary: ACTH, Gonadotrophins, Growth Hormone, Thyrotropin —The Adrenals: Corticosteroids, Aldosterone, Catecholamines—Serotonin—Renin—Thyroid Gland—Insulin—Sex hormones, Menstruation, Human Chorionic Gonadotrophin—Conclusion.

Pincus, 1943 (718), noted a daily variation in the excretion of 17-ketosteroids and thus indicated that a daily rhythm in adrenal function existed. It has since become apparent that circadian variations are a marked feature of the endocrine system. Adrenocorticotrophic hormone, follicle stimulating hormone, adrenal corticosteroids, catecholamines and testosterone have all been shown to exhibit circadian rhythms.

THE PITUITARY

ADRENOCORTICOTROPHIC HORMONE (ACTH)

The presence of a rhythmic variation in ACTH concentration in the plasma was shown by Liddle and his colleagues (288, 307, 308, 522, 672). In one experiment (672) blood samples were taken at 06.00 and 18.00 hours from a group of five male subjects to whom a constant 24-hour infusion of 0·9 per cent saline was given and also from six normal ambulant subjects and were analysed for ACTH. A fall in ACTH concentrations from 0·25mU per 100 ml plasma at 06.00 hours to 0·11 mU per 100 ml at 18·00 hours was found. In cortisol-deficient patients it has been found (307) that plasma ACTH concentrations still show normal daily variations, with ACTH levels much lower at 18.00 hours than at 06.00 hours. Even when the cortisol dosage was given in such a manner that the plasma corticosteroids were higher at night, the plasma ACTH levels were higher at 06.00 than at 18.00 hours in three out of the four patients tested. In patients recovering from long-term suppression of pituitary function, the plasma

ACTH levels also show a circadian variation with low values at 18.00 hours (308).

These variations in ACTH level are accompanied by parallel changes in plasma 17-hydroxycorticosteroids (17-OHCS) (672) and if ACTH is infused at constant rates over 24 hours, a similar relationship with plasma and urinary 17-OHCS concentrations exists up to 3 mU ACTH per 100 ml. The circadian rhythm of corticosteroid secretion (see below) thus appears to be under the influence of rhythmic changes in ACTH concentrations.

Variations in the sensitivity to stimulants and depressants of ACTH also occur. Metopirone (SU 4885) infused during the early morning produces a greater increase in ACTH release than at other times of the day (571). Variations in the suppressive effects of dexamethasone on ACTH release, as indicated by corticosteroid production, at different times during the day have also been demonstrated (673). 0·5 mg of dexamethasone given at midnight was found almost completely to halt corticosteroid production for 24 hours but the same dose at 08.00 hours or 16.00 hours merely suppressed corticosteroid production temporarily.

Further confirmation of the presence of a circadian rhythm in ACTH levels has been made by Demura *et al.*, 1966 (188), who, by using a sensitive immuno-assay technique, were able to estimate ACTH levels at 2-hourly intervals. Maximum values were observed in the early morning samples.

Schally *et al.*, 1960 (793), have suggested that the release of ACTH is regulated by a peptide, corticotrophin-releasing factor (CRF), from the hypothalamus. The production of CRF may, in turn, be regulated by higher centres in the brain and thus by the external environment. The anterior pituitary may, however, show variations in sensitivity (145). Corticotrophin-releasing factor (CRA 41) injected intravenously at 08.00 hours, 17.00 hours and 24.00 hours, produced a release of ACTH, as indicated by an increase in free 17-hydroxycorticosteroid concentrations in the plasma, which was greater at 17.00 hours and 24.00 hours than at 08.00 hours. As the adrenals are said to be less sensitive to ACTH at midnight (705) it is concluded that the anterior pituitary must be more sensitive to CRF. It has also been postulated (758) that two autonomous hypothalmic receptor sites both sensitive to corticosteroid feedback may be concerned in the control of ACTH release. CRF receptors are suggested as being responsible for rapid responses to corticosteroid variations and another controlling centre for rhythmic secretion.

PITUITARY GONADOTROPHINS

A circadian rhythm in follicle-stimulating hormone (FSH) concentration in man has been shown by Faiman and Ryan, 1967 (241), using a radioimmunoassay technique. Blood samples were obtained from twenty-three healthy men at 05.30 hours and 14.30 hours on two successive days. Samples were also collected at 4-hourly intervals from three subjects. The results showed mean concentrations of 0·201 mU/ml serum at 05·30 hours and 0·146 mU/ml serum at 14.30 hours in the large group. The 4-hourly estimations revealed highest concentrations at 05.00 hours with lowest values in the early afternoon. Luteinizing hormone (LH) was also assayed but no circadian rhythm was observed.

Prolactin

A circadian periodicity in the concentration of prolactin in the rat hypophysis has been noted (144) but in man this hormone is doubtfully distinct from growth hormone (549).

Melanophore hormone (MSH)

A circadian variation in pituitary melanophore hormone has been reported by Jores, 1938 (439), with a maximum in the early morning and a minimum about 18.00 hours.

THYROTROPIN (TSH)

A circadian rhythm in TSH concentrations with peak values between 02.00 and 03.00 hours has been claimed by some observers (81). Others have found no evidence of a circadian variation (681). The influence of a feedback mechanism from blood concentrations of thyroxine which in turn may be grossly altered by food intake (666) could possibly be responsible for these contradictory findings.

GROWTH HORMONE

A nocturnal rise in the secretion of growth hormone was noted by Quabbe *et al.*, 1966 (727). Hunter and his colleagues (411) attempted to trace the daily pattern of growth hormone (GH) concentrations in the plasma of adults. Samples were collected at approximately hourly intervals in six subjects during their waking hours. In only eight out of the 41 samples obtained was there sufficient GH for it to be measurable (0·5 μg/ml), and this represented merely the familiar relationship to meals and fasting (777, 778).

High values are, however, present in the plasma of children (412, 413) and in them a circadian variation has been found by Hunter and Rigal (414). Plasma samples were obtained from nine children aged 8-15 years, at hourly intervals during the day and 2-hourly at night. Seventy-two out of 142 samples collected had measurable quantities of GH. Consistently higher concentrations were seen at night in eight of the subjects. Hunter and Rigal suggest that GH may regulate growth by maintaining it continuously, especially during long intervals between meals, as at night time.

Glicks and Goldsmith, 1967 (290), later succeeded in carrying out a detailed investigation of plasma growth hormone levels in adults. They obtained blood specimens every 30 minutes over the 24 hours from six healthy young adults, two men and four women. The samples were taken through an indwelling needle. The subjects were confined either to bed or to a chair during the studies, apart from being permitted to go to the toilet.

Four different experimental regimes were followed. In one the subjects were kept fasting, in another half-hourly feeding of equicaloric liquid formula (Metrecal), totalling 1,800 calories for the women and 2,400 calories for the men, was permitted, in a third three liquid meals of an identical liquid formula were given to the same totals, and in the fourth a continuous infusion of 15 per cent glucose at a rate of 4ml/min for the men and 3ml/min for the women was maintained for 24 hours in the two men and two of the women and at a rate of 5 ml/min for 13 hours in the other two women. The blood samples were analysed for plasma growth hormone, insulin and glucose concentrations. Growth hormone levels were found to be higher at night than during the day under all four regimes with maximal values between 18.30 and 03.00 hours. A circadian rhythm in growth hormone concentrations was thus demonstrated and was shown to be independent of blood glucose concentrations and of the taking of meals.

POSTERIOR PITUITARY

Increased antidiuretic hormone (ADH) in morning samples of urine was noted by Schindl (797). Blood samples taken at

06.00, 12.00, 18.00 and 24.00 hours are claimed to have revealed higher levels of ADH at midnight than at the other times of the day in ten out of the twelve subjects examined by Zsoter and Sebok (944).

Recently Szczepanska *et al.* (853), have reported a sinusoidal curve in mean blood levels of ADH in 14 subjects living a nychthemeral existence from whom blood samples were obtained at 4-hourly intervals. Peak values were sometimes so low as to be undetectable but considerable individual variation occurred. The curves for the individual subjects were not shown but the authors state that most individuals showed a pattern similar to that of the mean. Goodwin, Jenner and Slater (301) have shown that the lowest excretion of antidiuretic hormone does not necessarily occur at night.

ADRENAL CORTEX

A rhythmic variation in adrenal function was suggested by Pincus, 1943 (718), who found regularly lower night values for 17-ketosteroids in urine specimens collected over a total of forty-eight 24-hour periods from a group of seven young men. Maximum values occurred in morning samples. Further investigations by Pincus and his co-workers (719, 772) confirmed the presence of a circadian rhythm in the excretion of 17-ketosteroids.

The development of a satisfactory quantitative method for measuring plasma 17-hydroxycorticosteroids (17-OHCS) by Nelson and Samuels, 1952 (667), enabled changes in plasma corticosteroid levels to be examined directly. A morning fall was quickly noted (786), and soon afterwards Bliss *et al.*, 1953 (79), demonstrated that this fall continued during the course of the day. They obtained three to five samples at 1- to 6-hour intervals from 08.00 hours to midnight from 18 subjects and also one to four samples at 1- to 4-hour intervals between 08.00 hours and 12.00 hours from 19 fasting subjects. Individual exceptions occurred but, in general, highest levels were seen in the 08.00-hour samples. Between then and midday there was a fall in both fasting and non-fasting subjects and lower levels persisted throughout the rest of the day, though again with some individual variations. Mean fasting plasma 17-OHCS concentrations in 267 estimations on a total of 120 subjects from whom samples were obtained between 08.00 and 08.30 hours was found to be 13 ± 6 μg/100 ml.

The occurrence of a morning peak in plasma corticosteroid levels followed by a decrease during the day was soon confirmed by a number of other observers (49, 257, 879).

Migeon *et al.*, 1956 (598), studied plasma 17-OHCS concentrations throughout the 24 hours and found that the variation in corticosteroid levels was circadian. Plasma samples were collected from twelve subjects at 08.00, 22.00 and again at 08.00 hours. Between 22.00 and 08.00 hours four of the subjects were awoken at 02.00 and 05.00 hours and plasma samples were taken by venepuncture. The other eight subjects had intravenous catheters inserted after the 22.00 hour samples had been obtained and thus samples could be obtained from them while they were asleep. One to three samples were thus obtained from these subjects at 02.00 hours, 04.00, 05.00 and 06.00 hours. Lowest mean plasma corticosteroid values were observed at 22.00 and 02.00 hours with no marked difference in trend between samples obtained by venepuncture and by catheter. At 04.00 hours an increase was seen to the highest value at 06.00 hours. Concentrations at the second 08.00 hours were similar to those seen in the samples collected at the first 08.00 hours.

A very thorough study of the circadian variation in plasma corticosteroids has been made by Perkoff *et al.*, 1959 (705). Plasma samples were obtained at 4-hourly intervals, 08.00 hours, 12.00 hours, 16.00 hours, 20.00 hours, 24.00 and 04.00 hours, from eleven normal subjects, five males and six females. The samples were analysed for 17-OHCS and, taken in conjunction with the results previously reported by the same laboratory (79, 598), provided very many plasma 17-OHCS concentrations over the twenty-four hours in fifty-one normal subjects. The majority had maximum values at 08·00 hours with a decrease thereafter to a minimum between 20.00 and 04.00 hours. Estimations of plasma 17-OHCS levels in five normal subjects on four successive days showed consistently lower levels at 16.00 than at 08.00 hours though the magnitude of the decrease varied.

Circadian variations in the urinary excretion of 17-hydroxycorticosteroids with consistently higher values during the day than at night and peak excretion in the morning samples have also been found (203, 494, 785).

Bartter *et al.*, 1962 (47), studied the urinary excretion of 17-OHCS and 17-KS in three groups of normal women. One group of seven college girls continued their normal daytime activities, and the other groups of five girls each remained within a hospital for the course of the investigation, which was carried out over 30-hour periods. Urine samples were obtained every 3 hours. As well as these 30-hour 'transverse' studies, 'longitudinal' studies over two weeks were carried out in hospital on the first group, in which plasma 17-OHCS were estimated at 07.00 hours and 19.00 daily. In the 'transverse' study a well marked circadian rhythm in the excretion of 17-OHCS and 17-KS was seen with minima at midnight and maximal values about midday. In the longitudinal study 07.00 plasma 17-OHCS values were consistently higher than those at 19.00 hours during the second week in hospital, but the pattern for the first seven days was quite irregular; with sampling at only 12-hour intervals, a rhythm could easily be overlooked if the chosen times were inappropriate.

In a number of recent studies highly specific methods have been used to estimate individual steroids. Vagnucci *et al.*, 1965 (881), using a sensitive double isotope technique, have been able to demonstrate a circadian variation in the excretion of unconjugated urinary cortisol both in healthy subjects and in three patients with Cushing's syndrome. Peak values in 10 out of the 12 normal subjects occurred in the morning, though two of the patients with Cushing's syndrome had afternoon peaks.

A circadian variation in both cortisol and corticosterone values in the plasma has been shown (706). In a group of six normal subjects, a 13-1 ratio of cortisol to corticosterone persisted throughout the 24 hours. A circadian rhythm in plasma corticosterone concentrations in three out of six subjects examined has also been demonstrated (572) using a fluorometric method.

It has also been shown (303, 773, 774) that the urinary metabolites tetrahydrocortisol (THF), 3α-allotetrahydrocortisol (ATHF) and tetrahydrocortisone (THE) have a circadian rhythm of excretion, and when both have been measured (205) renal excretion lags behind plasma concentration by about 3 hours.

Determination of the miscible pool of cortisol by using trace quantities of intravenous cortisol 4-C^{14}, has revealed that both the miscible pool and turnover rate of cortisol show a diurnal variation (710).

Since the changes in urinary excretion of corticoid closely follow the variations in plasma levels, the decrease in the latter during the day represents a fall in cortisol production (707).

Injection of labelled cortisol and examination of the specific activity of the cortisol metabolites (tetrahydrocortisone and tetrahydrocortisol) in the succeeding 24-hour urine, indicates that maximal cortisol production occurs in the late morning. The levels of both free and conjugated plasma 17-OHCS have also been found to exhibit a circadian rhythm with changes in the conjugated 17-OHCS lagging behind alterations in free 17-OHCS by 2 to 4 hours (118, 573).

A normal circadian rhythm in free plasma corticosteroids has been shown in women in the last month of pregnancy even in toxaemia, but in 33 per cent of women the conjugated plasma 17-OHCS rhythm was reversed at this time (701). Plasma corticosteroid levels are raised in the presence of high oestrogen levels as in pregnancy (151, 278, 596) and during oestrogen therapy (503, 601, 708, 770, 856, 905), but the normal circadian variation persists (151, 309, 770).

The circadian rhythms in plasma corticosteroids thus far reported were observed in subjects living a normal nychthemeral pattern of existence. Nightworkers have been reported as having a rhythm in plasma corticosteroid levels similar to that seen in subjects on a nychthemeral existence (598). The subjects studied were, however, either nightwatchmen or nurses and may not have been fully adapted to a nocturnal pattern of living. In people who are habitually on night work we have found (158) that adaptation of the plasma corticosteroid rhythm does occur with peak values seen in plasma samples obtained about the time of awakening (Fig. 3.1).

Sleep reversal experiments on nine subjects have been reported (705). The subjects were confined to a metabolic ward and were required to be in bed and asleep, if possible, from 08.00 to 16.00 hours and awake from 16.00 to 08.00 hours. Meal-times were altered appropriately and a considerable effort was made to exclude external time clues. The reversal period varied from five to ten days. In some subjects the 17-OHCS levels at 08.00 hours had fallen considerably by the fourth day, in others this did not occur until the eighth day. The concentrations at 16.00 hours changed in opposite direction so that the rhythm eventually became inverted. Thus in both the normal and inverted periodicity the rise in plasma corticosteroids occurred during the sleep period.

The effects of changes in sleep-activity routine on the rhythms of ketosteroid and ketogenic steroid excretion were investigated by Sharp *et al.*, 1961 (813). Four subjects were studied whilst living under standardized conditions of diet, activity and lighting in the perpetual daylight of the Arctic summer. After reversal of their

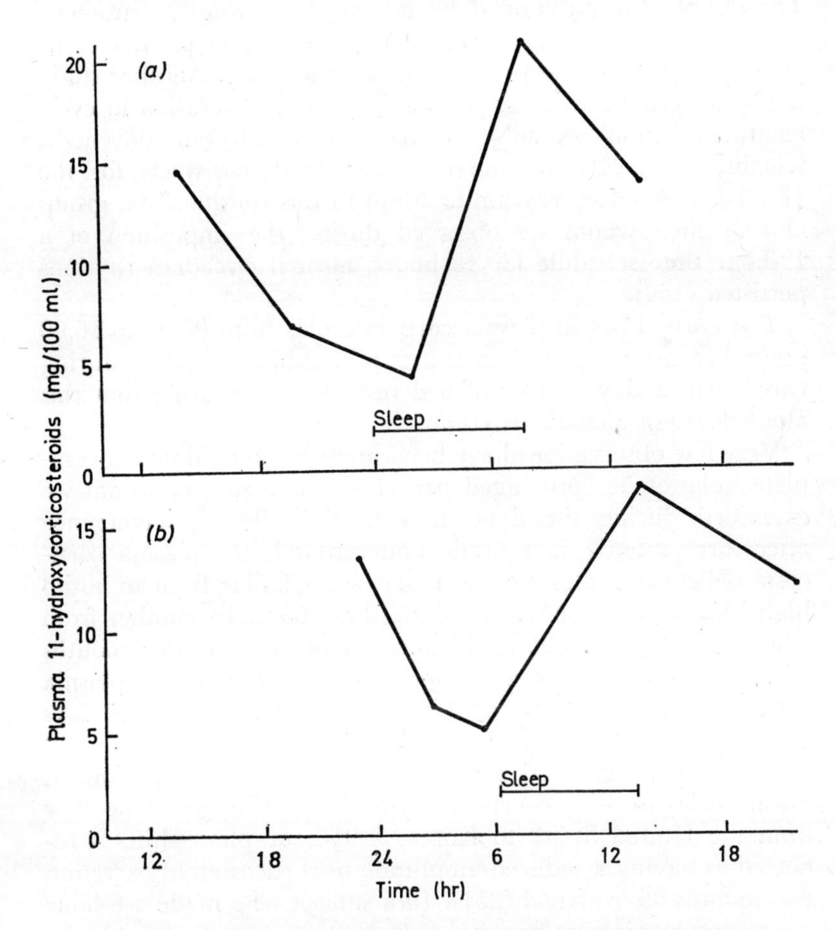

Fig. 3.1. Rhythm of plasma 11-hydroxycorticosteroid concentrations in (*a*) normal subjects, (*b*) nightworkers. (Conroy, R. T. W. L., 1969, Fig. 2, 160.)

sleep-activity regime, adaptation of the ketosteroid rhythm took two days and that of the ketogenic steroid rhythm eight days.

The imposition of abnormal time schedules such as a 12,- 19- or 33-hour rest-activity cycle can disturb the normal corticosteroid circadian rhythm. In subjects examined several days after such schedules had been imposed, maximal values were seen about the time of waking followed by a somewhat irregular decrease to minimal values shortly after the subjects had gone to sleep. The samples were obtained by means of in-dwelling catheters. Some of the individuals appeared to develop two plasma corticosteroid cycles per sleep/wakefulness cycle (686). Another study (826) suggested much less adaptability to an alteration in cycle length; a group of subjects lived on a 21-hour day/night schedule in the Arctic, and it took at least five weeks for the 17-OHCS excretory rhythm to adapt to this routine. In a group of four men whom we observed during the imposition of a 12-hour time schedule for 48 hours, normal circadian rhythms persisted (159).

The normal fall in plasma corticosteroids from 08.00 to 16.00 hours has been reported in a group of subjects studied during two fourteen day periods of bed rest (454) indicating that this alone does not alter the rhythm.

Very few observations have been made on individuals in complete isolation for prolonged periods. In one subject whom we examined during the days immediately following emergence after three months in solitude underground, the plasma corticosteroid concentrations showed no rhythm, falling from an initial high level and thereafter remaining low (607). In another from whom samples were obtained before emergence after four months isolation, whilst he was still ignorant of the time of day, plasma corticosteroid estimations at 4-hourly intervals over three days were suggestive of a 12-16-hour cycle (157).

Rapid transitions across time zones may result in prolonged disturbances in adrenocortical rhythmicity. A subject who flew from the United States to Korea, a 9½-hour phase-shift, is reported as having a reduced amplitude in corticosteroid excretion two months after arrival (251). In a subject who made a 6-hour phase-shift by air from England to Chicago we have found that the circadian rhythm in plasma corticosteroid levels was absent for at least eleven days and had not attained a normal amplitude 25 days after arrival (164).

Some evidence has already been produced that the steroid rhythm is associated with variations in ACTH liberation (188, 672). Some other possible influences on plasma steroid concentration must now be considered. De Moor *et al.*, 1962 (186), used gel filtration to study the binding of corticoids to plasma proteins, excluding the albumin binding, and estimated the bound corticoids both fluorimetrically and by measurement of radio-activity, sampling at various times of the day. They observed no circadian variation in plasma cortisol binding capacity, so the circadian rhythm in plasma corticosteroids does not seem to be due to changes in transcortin capacity.

To search for possible variations in ACTH sensitivity the effects of infusions of ACTH at various times, 08.00 to 14.00 hours, 16.00 to 22.00 hours and 24.00 to 08.00 hours in subjects on normal time routines have been observed (705). A dose of 40 units of ACTH was given over the 8 hours starting at midnight and 25 units in 6 hours at the other two times. The rise in plasma cortisol concentration was much less for infusions begun at 08.00 or 16.00 hours. The disappearance rate of infused cortisol, using a dose of 1 mg per kg. body weight was the same at all three times with half lives of 135 minutes at midnight, 132 minutes at 08.00 hours and 133 minutes at 12.00 hours; the lower plasma concentration attained with nocturnal infusions thus represents a lesser rate of production. Previous submaximal stimulation of the adrenal cortex with ACTH followed by a maximal ACTH test showed, however, that in such circumstances nocturnal adrenal production of 17-OHCS was at the same rate as during the day.

It appears then that variations in steroid removal from the blood, and in adrenal sensitivity to ACTH, are of minor importance, and that the rhythm in plasma corticosteroid concentrations is indeed due to changes in the rate of ACTH secretion.

As long as the ACTH concentration in the plasma does not exceed 3 mU per 100 ml, plasma and urinary 17-OHCS concentrations appear to be a rectilinear function of the logarithm of the plasma ACTH (Fig. 3.2). The closeness of the correlation has been shown (188) using a technique which allowed 2-hourly samples to be estimated. The immediate sensitivity to ACTH may, however, be influenced by past exposure to this hormone, causing the apparent variations in sensitivity (257, 674, 705).

That the alteration in ACTH secretion is in turn controlled by higher centres is implied from the finding that alterations in sleep and consciousness were closely associated with changes in the plasma corticosteroid rhythm. Support for this view comes from observations on subjects with brain damage (224, 395, 485, 486, 705). Out of 31 patients with various nonendocrine diseases (705) only in those with clinical states involving loss of consciousness or alteration of the sleep/wakefulness pattern was there significant loss of the corticosteroid rhythm.

In four out of seven patients with suprasellar tumours in which there was displacement of the hypothalamus (395), the circadian rhythm in 17-OHCS was lost. In these four patients there was a normal response of the adrenal cortex to ACTH stimulation and in four out of five patients tested ACTH secretion increased when

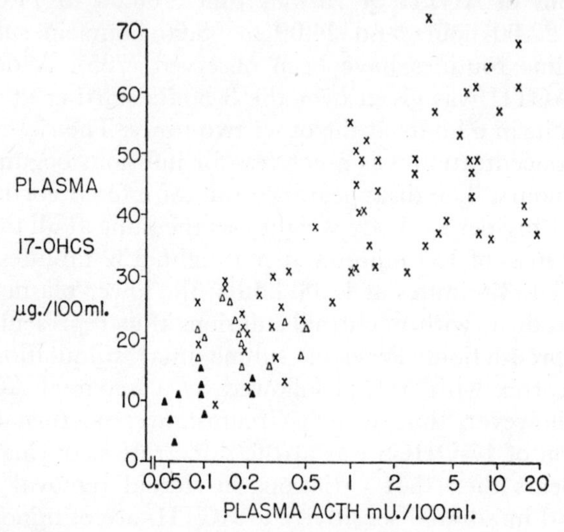

Fig. 3.2. Simultaneous plasma ACTH and 17-hydroxycorticosteroid concentrations. Each point represents a co-ordinate value for the two variables. The open triangles indicate the 6 a.m. values and the solid triangles the 6 p.m. values in untreated normal subjects. The values indicated by 'X' were obtained from normal subjects receiving 24-hour ACTH infusions. (Ney, R. L., Shimizu, N., Nicholson, W. E., Island, D. P. and Liddle, G. W., 1963, Fig. 3, 672.)

plasma corticosteroid level was reduced by an 11β-hydroxylase inhibitor (SU-4885). It is concluded that the loss of the circadian rhythm in plasma corticosteroids seen in some of these cases of suprasellar tumour was due to effects on hypothalamic structures either by a direct influence on the hypophysis or mediated by the brain stem.

Evidence that the temporal and pretectal areas of the brain play a part in the regulation of the circadian rhythm of plasma corticosteroids has been produced (485). Four patients with focal central nervous system disease not involving the sella and in whom there was no loss of consciousness were all found to have abnormal plasma 17-OHCS rhythms. Studies on a further 48 patients with focal disease of the central nervous system without disturbance of consciousness suggests that disturbances in the circadian rhythm of the plasma corticosteroids are associated particularly with central nervous system diseases affecting the pretectum, the temporal lobe or the hypothalamus and that alterations in the rhythm may occur in patients with normal sleep patterns (486).

An association between cerebral activity, rapid eye movements (REM) and the early morning increase in plasma corticosteroid levels has been suggested (913), but not confirmed, and if a subject is kept awake and thus has no REM his circadian rhythm in plasma corticosteroid values still occurs (674).

Whatever may be its own origin, adrenal rhythmicity probably plays a considerable part in the control of other periodicities. Its best documented influence is over the circadian rhythm in eosinophil levels, which will be considered in Chapter 5. As will be discussed in Chapter 4, its precise role in controlling the kidney is by no means clear.

ALDOSTERONE

A higher excretion of aldosterone by day than by night has been noted by a number of observers (46, 651, 653, 887, 931). Leutscher and Liberman, 1958 (553), estimated aldosterone concentrations in urine collected from three subjects over 8-hour periods from 07.00 hours to 15.00 hours, 15.00 hours to 23.00 hours, and 23.00 hours to 07.00 hours. Maximum values were seen in the samples collected between 07.00 and 15.00 hours and minimum values in the 8-hour period from 23.00 to 07.00 hours in all three subjects. A fourth individual showed a reversal of this pattern during a period of disturbed sleep and nocturnal activity with a return to the normal pattern on resuming the usual sleep/wakefulness and activity schedule. Bartter and his colleagues (45, 47), in a group of college girls, also showed a circadian rhythm in aldosterone excretion with low values during the night and a rise in the morning to a peak about noon, and also noted a secondary evening rise.

After three hours standing the excretion of aldosterone is increased but cortisol excretion is not raised (306) and immersion in water prevents this increase in aldosterone excretion. Posture, therefore, appears to play a major part in the control of aldosterone secretion and the apparent rhythmicity in aldosterone excretion seems to be mainly due to changes in posture (521, 552, 651, 653, 931) and may be mediated by variations in plasma renin concentration (930), though Bartter and his colleagues contest this view (45, 47) and report a patient with idiopathic oedema studied for two days when horizontal and two days when no longer confined to bed. The results analysed over 12-hour periods suggested that the rhythm was lost when the subject was recumbent but analysis over 3-hour intervals indicated that the magnitude of the rhythm was not lessened by the subject lying flat. Any loss in rhythmicity was considered merely to be a shift in time so that 12-hour collections represented equal amounts of aldosterone in the morning and night specimens.

ADRENAL MEDULLA

The urinary excretion of catecholamines in man has been found by von Euler and his colleagues (452, 896, 897) to be lower at night than during the day. Similarly plasma adrenaline has been found to be lower when asleep than after awakening both during the night and also in subjects asleep during the day (756). Recently Levi (506) in a detailed investigation of thirty-one subjects over a continuous period of three days under experimental conditions simulating wartime combat, has demonstrated a striking circadian rhythm in the excretion of adrenaline. Excretion reached a maximum in the early afternoon which was about three times the minimal excretion soon after midnight, although the subjects were maintained in a state of constant alertness and had no overt indication of the time. The rhythm in these experiments was clearly independent of external circadian influences and does not appear to reflect the suggested correlation between plasma adrenaline levels and mental activity (912). The subjects knew, though, that they were being fed every three hours and were thus able to calculate the time of day or night; some certainly appeared to do so (270). It has been suggested (936) that the adrenaline rhythms reflected the increased rate of conversion of noradrenaline to adrenaline under the influence of the increased morning production of cortisol (937). Noradrenaline excretion in these experiments also showed, however, a circadian rhythm, in phase about three hours ahead of that of adrenaline and of much lower amplitude (Fig. 3.3), so the adrenal medulla as well as the cortex may show circadian rhythmicity.

SEROTONIN

A circadian rhythm in 5-hydroxytryptamine concentrations in human blood has been reported by Halberg and his colleagues (61, 330). From a group of five healthy adult male subjects and eleven male patients with mental defects, mainly Down's disease, blood samples were obtained at 4-hourly intervals. The crest lay between 04.00 and 14.00 hours but the rhythm had a very small amplitude, 1·7 to 2·2 μg/100 ml against an overall sample mean

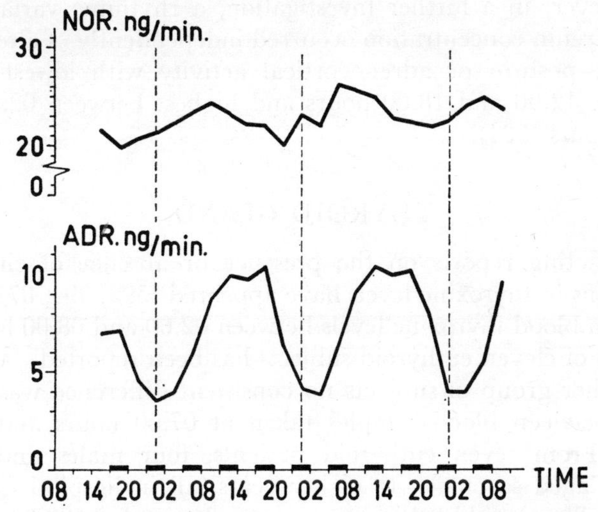

FIG. 3.3. Urinary excretion of noradrenaline (NOR) and of adrenaline (ADR) in 63 subjects awake and active for 3 days. (Fröberg, J. and Levi, L., unpublished data.)

of 19 μg of 5-hydroxytryptamine per 100 ml. By sophisticated statistical analysis using computer-based cosinor tests (347) a significant circadian rhythm was, however, detected.

PLASMA RENIN

Plasma renin concentrations were shown by Brown *et al.*, 1966 (119), to vary over the 24 hours in samples taken at 6-hourly intervals. A 31 per cent increase in values between 04.00 and 10.00 hours was found. Posture, exercise and diet were uncontrolled, and it is suggested (302) that increased renin secretion following standing may be influenced by release of catecholamines. By compression of the carotid sinus or by bandaging the legs of subjects before allowing them to stand, the normal posturally induced rise in plasma renin activity may be abolished. It is concluded that pooling of blood in the upright posture produces reflex sympathetic nervous activity and release of catecholamines and that this may induce renal arteriolar constriction and increased secretion of renin. These observations recall the effects, described above, of posture upon aldosterone, whose production is stimulated by renin.

However, in a further investigation, a rhythmic variation in plasma renin concentration occurred independently of alterations in diet, posture or adrenocortical activity with lowest values between 12.00 and 18.00 hours and highest between 02.00 and 08.00 hours (304).

THYROID GLAND

Conflicting reports on the presence or absence of circadian variations in thyroxine levels have appeared (592, 794, 871, 902). A rise in blood thyroxine levels between 02.00 and 08.00 hours in a group of eleven euthyroid subjects has been reported (902) but in another group of subjects no consistent difference was found (871) between blood samples taken at 07.00 hours and 17.00 hours. From seven euthyroid patients, four males and three females aged seventeen to eighty years, all in hospital for non-thyroid illnesses, blood samples were obtained at 08.00, 13.30, 17.30, and 03.00 hours and were analysed for protein-bound iodine (794); no marked fluctuations were found.

A rhythm has been shown, however, in a group of fifteen euthyroid students living a normal existence, from whom samples were collected every 4 hours except during the night (592). Total serum iodine showed no significant changes, but mean P.B.I. concentrations was minimal at 08.00, hardly changed until 16.00, and then rose to a maximum at 24.00; this was also the time of the maximum in 13 of the 15 individuals, the other two attaining their maximum at 20.00 hours. The circadian pattern was thus fairly consistent although the absolute levels differed considerably. Other observers have generally failed to sample at the time when the maximum concentration was found in this study, but the differences may also be due to the different populations sampled;

healthy students and hospital patients represent highly selected fragments of the population at large. Furthermore thyroxine levels are dependent not only on its secretion rate but also on its rate of removal by the two principal channels, i.e. tissue utilization and biliary excretion and partial subsequent reabsorption. These factors cannot be controlled in humans but experiments on rats (666) give some slight indication that independently of food intake, tissue utilization by night is greater than by day.

INSULIN

Cyclical variations in plasma insulin levels have been noted (498) in a group of 16 non-fasting subjects, in whom higher immunoreactive insulin (IRI) values generally occurred at night. A series of studies (266) has confirmed and extended this finding. Two groups of normal subjects were kept fasting, five subjects for 3 days and six subjects for 4 days. Blood samples were obtained in the morning at 07.00-08.00 hours, and in the afternoon at 15.00-16.00 hours, and were analysed for IRI and glucose. IRI levels were significantly higher in the morning samples throughout and the ratio for IRI/glucose was greater in the morning than in the afternoon. No rhythm in plasma glucose levels was discerned. As the authors suggest, insulinogenesis might occur secondarily to cyclical changes in plasma glucose levels below the limits of accurate analysis, or, if there are no differences in morning and afternoon glucose levels in normal subjects, more insulin may be required to maintain normoglycaemia in the morning.

SEX HORMONES

ANDROGENS

Migeon *et al.* (597) have recorded a circadian variation in dehydroepiandrosterone (DHEA) levels in plasma as have Eik-Nes *et al.* (225), with significantly lower values at 08.00 hours than at any other time in the 24 hours. Lower plasma androsterone levels were also seen at this time (597) but most reports of circadian variation in plasma testosterone levels (216, 461, 481, 532 ,757, 834, 835) show a different timing.

Blood samples obtained from five healthy young adults aged twenty to forty-two years at 07.00 hours, 13.00 hours, 19.00 hours and 01.00 hours showed in all subjects a markedly low plasma testosterone level at 01.00 hours (Fig. 3.4). Concentrations at 07.00

hours and 13.00 hours did not differ significantly from one another. Similarly, samples examined at 09.00 and 17.00 hours have not been found to vary (481). A more extensive investigation (757) has confirmed the occurrence of a daily variation in plasma testosterone levels. Samples were taken at 4-hourly intervals from 08.00 to 24.00 hours from 20 healthy male subjects aged twenty-one to twenty-eight years. Significantly ($p < 0{\cdot}01$) higher mean plasma values were found at 08·00 hours than at other hours.

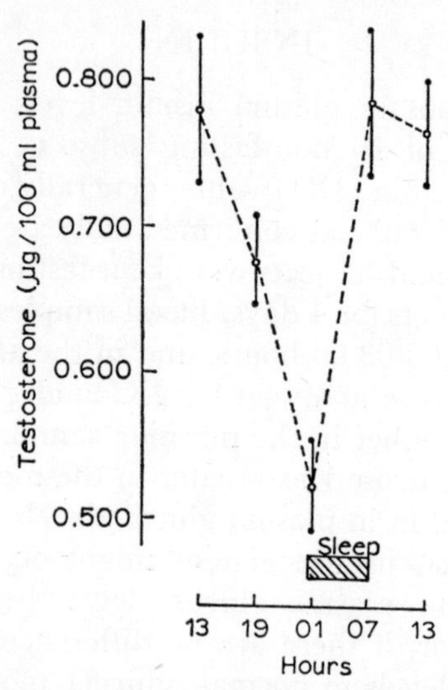

FIG. 3.4. Variations in plasma testosterone concentration during the 24 hours. (Dray, F., Reinberg, A. and Sebaoun, J., 1965, 216.)

Since testosterone is produced by both the testes and adrenals, the rhythm is not necessarily due to an early morning rise in pituitary gonadotrophin secretion, but could, as the authors point out, be caused by a fall in adrenal testosterone secretion in the afternoon and night.

The metabolic clearance rate (MCR) and plasma concentrations of plasma testosterone have been measured (834) in two normal young men and two normal young women confined to bed in hospital. Plasma samples were taken at ninety minute intervals (16 specimens) and were pooled for each subject and analysed for testosterone. Plasma samples obtained at 4-hourly intervals (09.00, 13.00, 17.00, 22.00, 02.00 and 06.00 hours) were also analysed individually. The concentration found in the pooled specimens was taken to indicate the daily mean and was compared with the concentrations in the individual 4-hourly specimens. A marked circadian variation in plasma testosterone levels was seen in both men, with highest values in the samples taken at 09.00 hours. Minimum concentrations were seen at 22.00 hours in one subject and at 02.00 hours in the other. The plasma testosterone level in the pool, which was taken to represent a close approximation of the daily mean, was 22 per cent lower than the value at 09.00 hours in both subjects. The female subjects did not show any significant circadian variation in testosterone levels, in agreement with an earlier report (497). The metabolic clearance rate was measured by a constant infusion technique in two other normal male subjects and compared with plasma testosterone levels. No significant variation in MCR was seen during the course of the day. The fall observed in the concentration of testosterone must therefore be due to a decreased production of testosterone and not to an alteration in MCR.

PROGESTERONE

Plasma progesterone levels have been measured by gas chromatography at 06.00 hours, 11.00 hours and 18.00 hours in three patients in early pregnancy (555). The results suggest a rhythmical variation in plasma progesterone. Highest values were seen at 06.00 hours on the first day of admission to hospital and 11.00 hours thereafter. A 24-hour rhythm in the excretion of pregnanediol in pregnancy has also been reported (561); in urine samples collected over 6-hour periods maximum concentrations were found in the 00.00 and 06.00 hours specimens.

MENSTRUATION

Malek and his colleagues (564) analysed the time of onset of menstruation in 7,420 menstruations in 810 girls between 14- and 18-years-old. During the night the onset of menstruation was stated to be much less frequent than during the day. The most common time of onset was between 04.00 and 12.00 and the least frequent between 20.00 and 04.00 hours. A postural basis for this variation would seem highly likely.

HUMAN CHORIONIC GONADOTROPHIN (HCG)

Serum HCG levels estimated by the rat ovarian hyperaemia test in 16 pregnant women at various times during the day showed no evidence of a circadian pattern (91). Urinary HCG estimated immunologically has, however, been reported (34) as showing peak excretion between 07.00 and 09.00 hours in a group of 24 patients. In a group of six pregnant patients in whom 4-hourly serum and urine titres were examined by another observer (861), serum levels were found to be constant as in ten subjects in another investigation (628). Considerable random variations in urine titres were seen, but the urine flow is not recorded. The very high serum concentrations in women with hydatidiform mole are likewise very constant throughout the 24 hours, while the urinary concentrations vary irregularly (862).

CONCLUSION

Circadian variations thus occur in a large number of endocrine functions. In relatively few, however, is there good evidence that endogenous rhythmicity exists and that the rhythm is not simply due to habit, posture, or some other exogenous cause. The most carefully authenticated is the rhythm in adrenal corticosteroids. Its relationship to other rhythms has been widely studied both in nychthemeral existence and under a variety of abnormal schedules, but it is not yet clear whether its close association with many other manifestations of rhythmicity indicates a common cause, or implies that the adrenals control the other manifestations.

Chapter 4

The Kidney

Urine Flow—Electrolytes—Acid and Alkali—Phosphate—Calcium and Magnesium—Glomerular Filtration—Miscellaneous Solutes—Cause of the Rhythms—Adrenal Cortex—Influences on Sodium Excretion.

URINE FLOW

It is usual to sleep for 8 hours or so, uninterrupted by any need to pass urine. Reflective men must have realised that this implied a low urine flow, though the earliest record of this observation we have been able to discover was by Schweig in 1843 (804). Vogel in 1854 (890), made the same observation and supposed it was due to the low intake of fluid by night, a conclusion which would have been accepted, one might presume, by any earlier observer of the phenomenon. In the nineteenth century, however, speculation was giving way to experiment and in 1860 Roberts (769) showed that this low flow by night was not simply due to the low fluid intake, a conclusion which soon found further support (493, 729). The idea that these variations of urine flow might be due to some intrinsically rhythmic process was overtly expressed in 1927 by Völker (893), who studied subjects under carefully controlled conditions in a dark room on a rigid daily schedule with controlled fluid intake and weighed meals. The subjects were allowed to sleep, but he demonstrated that if they altered their sleeping habits the habitual urinary flow rhythm might remain unaltered. He also seems to be the first to have shown that the rhythmic alternation of light and darkness was more important than their relative duration, since the behaviour of the kidney, and of other physiological functions which he also studied, was much the same in midsummer in the Arctic or in midwinter in Hamburg. Jores (435) made similar critical controlled observations and came to the same conclusion. Hart and Verney (356) observed a spontaneous increase of urine flow in the morning in subjects who remained recumbent and fasting and in conditions very carefully controlled to avoid any disturbances of bodily water balance. They ascribed the changes to alterations in the level of secretion of posterior pituitary antidiuretic hormone, though this explanation is open to doubt.

That urine flow shows an endogenous rhythm is now abundantly established. Urine flow is not, however, in itself a primary physiological variable but is rather the resultant of a number of other physiological processes: the rate of glomerular filtration and of tubular reabsorption, the reabsorption of various individual solutes, and the secretion of the posterior pituitary antidiuretic hormone (ADH) and of other hormones, to name a few. There is little evidence for any regularly rhythmic secretion of ADH, see Chapter 3: an early claim (944) has been refuted (296, 301) and reasserted (853). Moreover the sparse urine produced at night has usually a lower electrolyte concentration than that produced during the day (613), the reverse of what one would expect if the low flow were due to an increased secretion of ADH. This suggests that the true explanation of the changing urine flows lies mainly in changing rate of excretion of the various urinary solids.

The close association of urinary volume with electrolyte excretion is well shown in Fig. 4.1. This would seem to justify the contention that when other circumstances are held reasonably constant, the circadian rhythm of urine flow is osmotically determined, though of course in subjects following patterns of existence of their own choosing, many other influences may enter in. Sophisticated mathematical techniques applied to an extensive series of data on a single subject (323, 335, 340) demonstrated a weekly rhythm; this was apparently due to the regular habit of the subject of drinking a bottle of beer on Sunday.

Sodium, potassium and chloride account for a large part of the osmolar content of the urine and are thus likely to exert a major influence upon urine flow, so their excretory pattern will next be considered.

ELECTROLYTES

Sodium, potassium and chloride are all excreted in rather small amounts at night and in increasing amounts in the morning, reaching a peak commonly around the same time as one another in the morning or early afternoon. The detailed pattern varies somewhat with the activity of the subjects, but it has been clearly demonstrated that this rhythm is truly endogenous and independent of the various factors in the environment that might influence it. Simpson controlled the more obvious influences of food intake and light and darkness though he allowed his subjects to

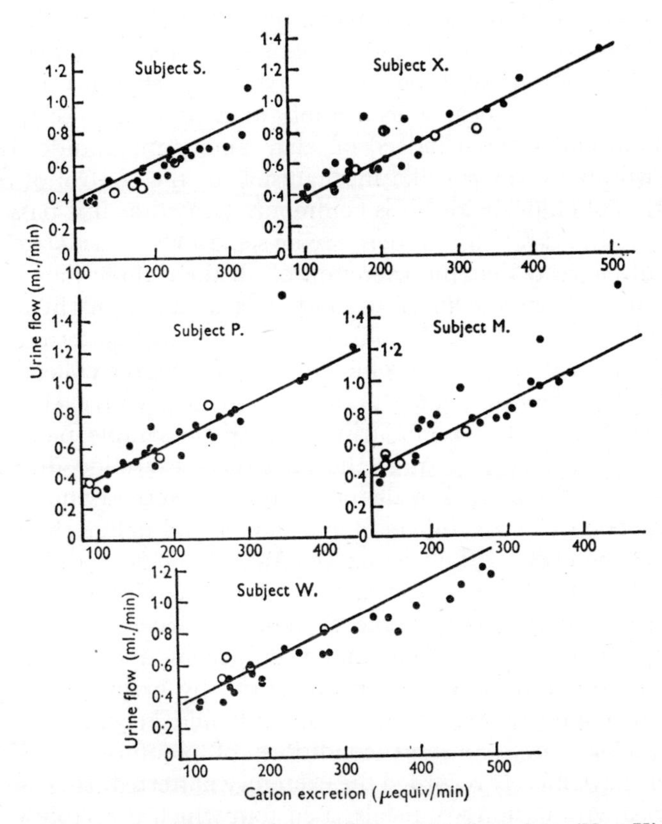

Fig. 4.1. Dependence of urine flow upon electrolyte (Na+K) excretion in five subjects living on a 12-hour day; ● awake, ○ asleep. The regression line is calculated from the pooled data from all five subjects. (From Mills, J. N. and Stanbury, S. W., 1952, **117**, 30.)

sleep (823); he claimed that this influenced chloride excretion, which in the morning was lower in subjects who went to sleep again, and by night was higher in those who stayed awake. The data on this point are rather scanty and not altogether convincing but Norn (677) makes the same claim, that at any time of day electrolyte excretion is lower during sleep than if the subject is awake.

Most of the earlier observations upon electrolyte excretory rhythms (42, 51, 465, 493, 823) confined themselves to chloride, although occasional observations upon sodium and potassium were made by the very tedious chemical methods then available (566, 677). Only with the advent of flame photometers were extensive studies of sodium and potassium excretory rhythms feasible. Chloride excretion nearly always follows a pattern close to that of sodium and is often indeed of fairly similar magnitude. Little is known about any specific mechanisms for the control of renal excretion of chloride and it is commonly treated as if it passively follows sodium; it will not therefore be separately considered. It is generally agreed that the excretion of all these three ions is low at night, rises in the morning and falls again at night at the customary time of sleep. Accounts of the excretory pattern during the day differ in detail and some are distorted by the collection of urine only over rather long periods of 4-6 hours or even more. Most workers collect night urine as a single sample passed on waking, so little is known of the pattern of excretion during a night's sleep. Some workers define the time of excretory maximum by inspection whereas others fit sine curves and define the maximum as the crest of the sine curve. Apart from such differences in the collection and presentation of data, the circumstances of study have varied widely. Some, even of the earlier workers (566) have taken care to exclude major influences upon electrolyte excretion other than a circadian rhythm, by giving water and meals in regularly spaced small amounts and keeping the subjects under fairly constant conditions of recumbency. Other workers have merely recorded the excretory pattern during normal existence with customary meals, a custom which probably varies widely between different people; the ingestion of caffeine-containing beverages, for instance, is in some subjects a potent stimulus to sodium excretion and may completely distort any circadian rhythm. Sodium excretion appears to be more susceptible to interference by uncontrolled influences than does excretion of potassium and consequently it is less regular or predictable. Potassium excretory peaks however have been recorded as early as between 09.00 and 11.00 hours (419, 614) or as late as 18.00 to 19.00 hours (608), though most workers (515, 569, 570, 807, 841) find a maximum just before or just after midday. Those who have made the comparison (214, 370, 623) have observed that the excretory peaks on a normal routine are often rather later than under constant conditions, and may even be postponed until the evening; this may merely reflect dietary habits, such as drinking coffee or cocoa. A similar conclusion can be drawn from experiments on an abnormal time schedule (426, see also Chapter 9).

It has been claimed (282) that the potassium excretory rhythm is affected both by geographical location and by time of year. Groups of six to eight subjects were studied in Paris in the autumn and the spring, in Copenhagen in the autumn, and again in Paris in the autumn. Urine collection was in 4-hourly periods and maximum excretion was observed between 08.00 and 12.00 hours in the autumn in Paris, between 12.00 and 16.00 hours in the autumn in Copenhagen and between 16.00 and 20.00 hours in the spring in Paris. The subjects were apparently living a normal existence and it is impossible from the published data to determine how far these differences may have resulted from changes in habit or diet. No systematic study appears to have been conducted concerning possible influences of time of year or geographical location such as suggested in this paper, apart from the extreme conditions of the Arctic considered in Chapter 8.

Though numerous attempts have been made to remove the major interfering influences upon sodium and potassium excretion in order to reveal the uncomplicated action of the supposed circadian rhythmic clock, it is doubtful whether such attempts have ever been entirely successful. The nearest approach was probably in experiments (614, 615) in which four subjects each spent 24 hours continuously working in the laboratory, not eating and taking either 100 ml water or 100 ml 10 per cent glucose every hour; electrolyte excretion showed considerable irregularities in the first six hours, perhaps ascribable to the uncontrolled meals or activities before the beginning of the experiment, but in the last eighteen hours electrolyte outputs followed a smooth sinusoidal course (Fig. 4.2). These experiments however still fail to satisfy the most stringent criteria for complete isolation from 24 hour influences since, although lights were on continuously in the laboratory, this was quiet at night whereas during the day daylight entered through the windows and other people were around. It has been suggested (21, 22, 338) that such social influences can be an important Zeitgeber in man. In animal experiments indeed it has seemed (116) that even the regular visit of an attendant to feed the animals and clean the cages is quite sufficient to impart a 24-hour rhythm to the animals' physiological behaviour. It is more practicable in man to study only part of a cycle; for instance, one may remain recumbent

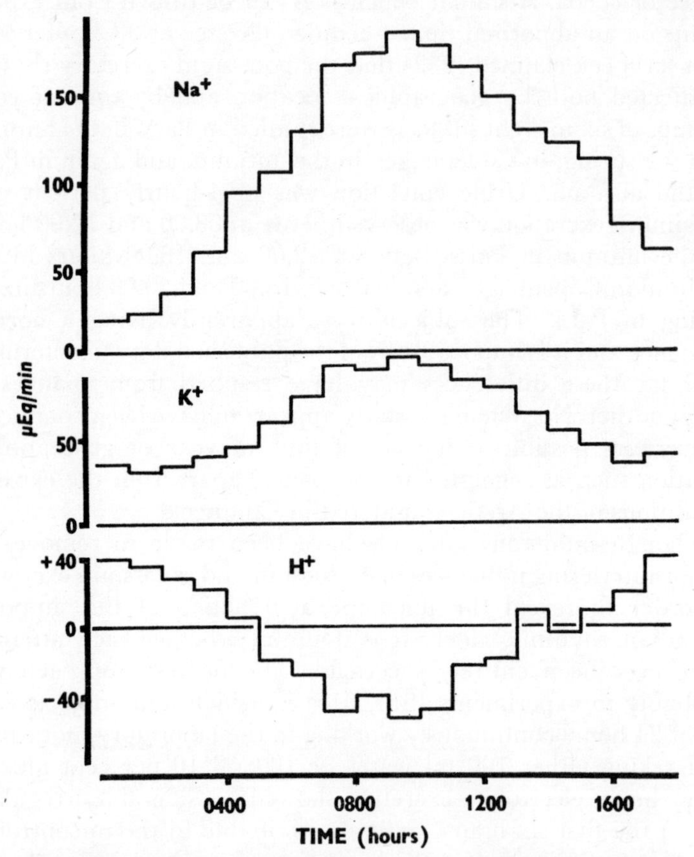

FIG. 4.2. Sodium, potassium and hydrion (titratable acid + ammonium − bicarbonate) excretion in a subject continuously engaged in laboratory work, and taking small identical liquid meals every hour. (In part from Mills, J. N. and Stanbury, S. W., 1954, 614.)

and fasting from the time of waking in the morning until the early afternoon. Observations in such conditions (419) show a fairly reproducible pattern on different occasions on the same subject, without gross differences between one subject and another (Fig. 4.3). Even complete isolation from all rhythms of social contact or of climatic factors, as in a deep bunker (32), or a cave (607), is not satisfactory for study of intrinsic renal rhythmicity since subjects under these conditions follow some sort of pattern of meals, sleep, activity and so forth, all of which may influence the kidneys.

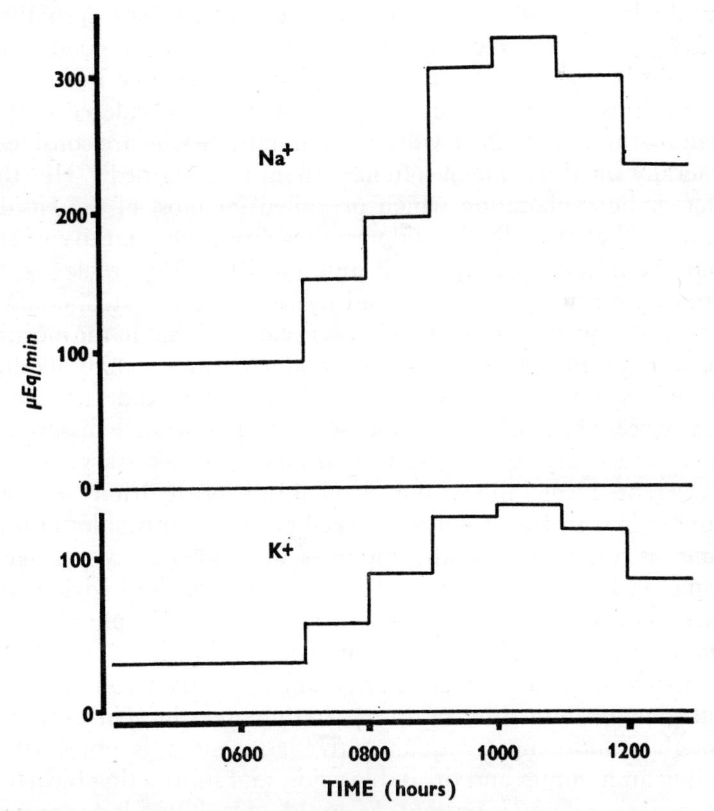

FIG. 4.3. Sodium and potassium excretion, in a subject remaining recumbent and fasting after waking in the morning. Mean of 5 days. (In part from Imrie, M. J., Mills, J. N and Williamson, K. S., 1963, 419.)

Since in man constancy of all potential influences is unattainable for a sufficient period, the strict proof of an endogenous rhythm must rest upon its persistence in subjects who are not following usual nychthemeral habits. Observations on subjects subjected to a sudden phase-shift, or living on a 'day' of abnormal length, are described in Chapter 9.

ACID AND ALKALI

The earliest record that the acid urine produced at night is often replaced by alkaline urine about noon seems to be that of Dr. Andrews of Belfast who described his findings verbally to Bence-Jones (431). He claimed that of 15 students in good health two thirds voided at noon urine which was alkaline in reaction. Jones, drawing on his own further experience, speculated that 'in London this alkalescence will be found in those who are considered generally healthy much oftener than is imagined'. He then offered the explanation which prevailed for most of the ensuing century, that this alkaline tide resulted from the quantity of acid poured out into the stomach. Brunton in 1933 (124) reviewed the considerable literature which had by then accumulated, confirming that the scanty urine at night was acid, and that in the morning the acidity diminished and the urine often became frankly alkaline. He points out that pH alone is an inadequate guide to urinary acid secretion, and that some sort of quantitative assessment based upon, among other things, urine flow is necessary.

Current ideas about the mechanism of hydrion secretion suppose that its transfer from the cell of the renal tubules into the lumen is followed by reabsorption of bicarbonate, by passage of ammonia into the lumen, and by buffering of the hydrion by a variety of buffers within the lumen, of which phosphate is quantitatively the most important (606, 934).

Change of excretion of ammonium or of titratable acid will thus give an indication of directional changes in acid excretion, though a more nearly quantitative assessment is obtained by adding ammonium and titratable acids, and subtracting bicarbonate excretion if any is present. Even this calculation is not entirely adequate since the bicarbonate reabsorbed can only be calculated if one knows the filtered load, obtained by multiplying the volume of fluid filtered at the glomerulus by the plasma bicarbonate concentration. The determination of total hydrion secretion by the tubular cells is therefore a difficult performance and has not often been attempted over a 24-hour span. In most circumstances, however, it is presumed that neither GFR nor plasma bicarbonate concentration changes very much so that the calculation given above yields quantitative information about changes in hydrion secretion. In considering the further literature, then, changes in any one component will be accepted as indications of the directional change of the whole. Further accounts of the nocturnal acidity of the urine may be found (57, 428, 465) and this has been shown to be independent of food (51, 149, 254, 505, 841) and of sleep (135, 149, 455, 487, 823). There is usually a close reciprocal relationship between excretion of potassium and of hydrion when these are varying circadianly (614, Fig. 4.2) and this relationship may persist on an abnormal time regime (607). It is thus surprising to find one account in which the hydrion excretion is said to be low by night and only to rise by about midday (47). This is reported without any comment that it is at variance with the great mass of other determinations, and even with a suggested explanation which is almost exactly the reverse of the usual suggestions. In a second atypical account (214) it has been claimed that hydrion excretion is better correlated with sodium than with potassium excretion. There appears, indeed, to be considerable individual variation in the pattern of renal acid excretion; in a study of urinary pH on 18 subjects, living on usual nychthemeral habits (228), the characteristic pattern was the familiar nocturnal acidity, followed by a morning alkaline tide; in some subjects, however, the morning peak was absent, in some the urine remained acid throughout the 24 hours, and one subject produced alkaline urine only at night. Whether the anomalies were persistent characteristics of the individuals, and associated with any peculiarities of habit, is not recorded, but it is known that the dietary habits on the previous day, and even trifling amounts of exercise on the day of observation, can profoundly affect the acid and alkali excretion by the kidney (42).

PHOSPHATE

One of the factors determining renal excretion of acid is the amount of buffer available, and the principal such buffer is phosphate. However acid and phosphate excretion are often out of phase with one another. Phosphate excretion usually falls for some hours after waking, and when samples are taken during the night it has usually been observed that the fall was already proceeding during sleep. The time of the peak excretion seems to be somewhat variable, as indicated in the assembled data from three groups of workers (608). This morning fall in excretion of phosphate has been very generally recorded (47, 51, 104, 135, 250,

482, 608, 623, 841) but the minimum may occur so early that excretion is rising during most of the morning (297). It adapts quickly to changes of routine (135), and indeed in a subject changing back to day-work after a spell of four weeks night-work (617) the phosphate rhythm adapted immediately, excretion falling after each period of sleep. The same has been observed in subjects living in caves and following unusual or quite irregular habits of sleep (607, 609). Only one account in conflict with this picture has been discovered, a claim that the phosphate excretory rhythm is unaltered in men working on a night shift (482). The pattern is also said to be disrupted by a few days' sleeplessness (487).

The phosphate excretory rhythm adapts itself so readily to every change in habit that the evidence for an endogenous rhythm is not strong. It rests chiefly upon the very regular oscillatory behaviour of phosphate excretion in subjects spending 24 hours or more under constant conditions of rest or activity (608), and upon its immediate reappearance after it has been disrupted for three days (622).

Phosphate excretion is determined by a Tm mechanism, whereby the proximal convoluted tubule reabsorbs up to its maximal ability (15); thus the rate of excretion will be determined by plasma concentration, by glomerular filtration rate and by the reabsorptive ability of the tubule. The circadian rhythm in subjects maintained under reasonably constant conditions, as well as the sharp fall in excretion usually seen in the early morning over shorter periods of study, are paralleled by changes in plasma phosphate concentration (69, 136, 608, 615, 840, 841, 915) (Fig. 4.4) but some other variations in excretion involve altered tubular activity.

The cause of the usual morning fall in concentration of phosphate in the plasma has never been fully investigated. It is tempting to ascribe it to the early morning secretion of cortisol, whose plasma concentration changes in the opposite direction to that of phosphate (Chapter 3). Cortisol causes a sharp fall in plasma phosphate (616), by promoting its uptake by skeletal muscle (619), and can also diminish phosphate excretion without altering either G.F.R. or plasma phosphate concentration (297). In seeming contradiction to this simple hypothesis is the general experience that the cortisol rhythm is very stable in the face of changes of habit and routine, whilst the morning fall in phosphate excretion adapts immediately to changes of habit. When cortisol secretion is suppressed by prednisolone (419) this morning fall in phosphate excretion persists, and when cortisol synthesis is blocked by metopirone the morning fall in phosphate excretion is actually somewhat greater than in control experiments. Unfortunately

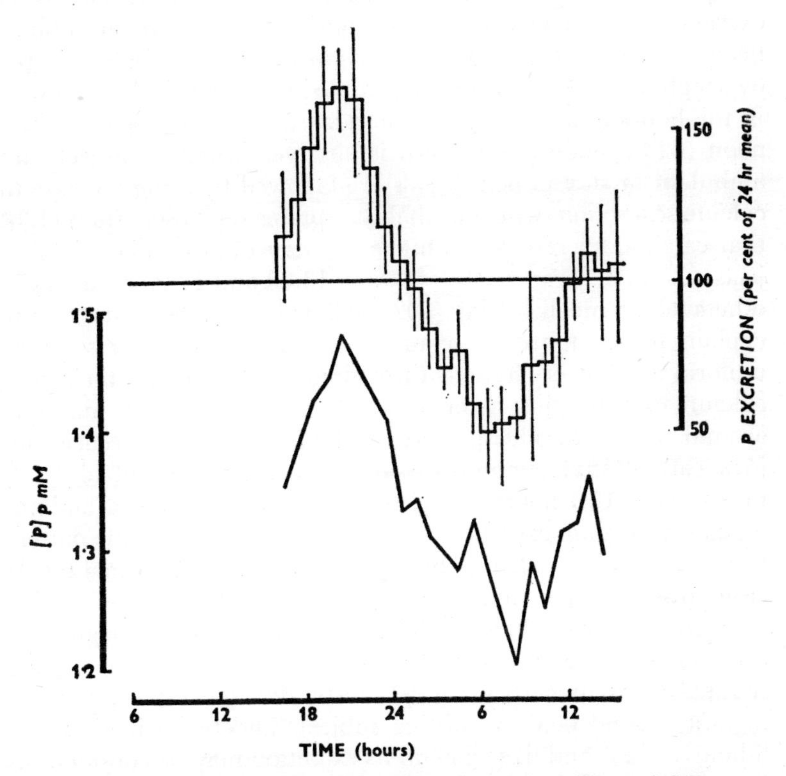

FIG. 4.4. Above, mean urinary phosphate excretion, and S.D.; below, mean plasma phosphate concentration; in subjects awake and active and with identical hourly food and fluid intake. (From Mills, J. N., 1966, Fig. 1, 608.)

none of these observations included measurements of plasma phosphate concentration, and since other factors, such as glomerular filtration, can affect phosphate excretion, it is not certain that the excretory data reflect changes in plasma concentration. Certainly in hyperthyroid subjects morning phosphate excretion is less and plasma phosphate higher than in euthyroid subjects (627).

CALCIUM, MAGNESIUM AND OTHER METALS

There is still some conflict of evidence about the circadian rhythm in calcium excretion, although most workers agree that the plasma concentration remains very constant, variations in excretion presumably resulting from changes in renal tubular behaviour. In fasting subjects, calcium excretion is highest in the overnight sample or in the first sample in the morning, and lowest in the hours immediately after midday (370), or in the late afternoon (623). The same pattern is observed whether subjects are ambulant or stay in bed. Meals are followed by a big increase in calcium excretion, which probably explains the observation (136) that calcium excretion was higher by day than by night in subjects who took three meals, at 09.00, 13.00 and 17.00 hours, as in others taking meals by day (102, 482, 623, 915). More difficult to explain however are the claims (69, 623) that subjects taking uniform meals every 3 hours throughout the 24 showed their peak calcium excretion just before or after midday, and their minimum around midnight. There appear to be individual differences, for in a careful study upon two subjects on a routine of identical meals every two hours (550), one showed a minimum and the other a maximum excretion around midday and the same pattern was seen on a low-calcium diet. Exercise, as well as normal meals, easily disturbed the rhythmicity.

There is similar disagreement about the pattern of magnesium excretion, although fewer studies have been performed. It is reputed to be higher by night than by day (205, 370), or to be highest around 06.00 in fasting subjects and 08.00 in those fed 8-hourly (623) and like calcium its excretion may be considerably increased after meals (370), and may not show a clear rhythm (915). If the 24-hour urinary output is divided into two aliquots by collecting at 08.00 and 20.00 hours, magnesium and calcium excretion are said to be lower in the night sample (102), but this is a coarse procedure for defining rhythms. The pattern of excretion of magnesium and calcium has recently been studied under conditions of altered time schedules together with the sodium and potassium excretory rhythms. In subjects living a 22-hour day all these electrolytes were shown, by fitting cosine curves to the data, to have a 24-hour component, indicating an endogenous rhythm, as well as a larger 22-hour component, indicating a major influence of the subjects' meals and other habits (426).

In subjects exposed to lead or mercury in the course of their work, excretion of these metals shows a circadian rhythm (640, 641) which is not determined by the hours during which they are exposed to these metals, since the pattern of excretion is the same in those working on day or night shifts. The cause of these rhythms is at present completely unknown.

GLOMERULAR FILTRATION

A rhythm in glomerular filtration rate has been demonstrated by measurements of inulin clearance, and in all such studies glomerular filtration was higher by day than by night (827, 841, 882, 915, 916). There is, however, much individual variation, and in each series there were subjects whose peak diurnal clearance exceeded the nocturnal minimum by more than 20 per cent, though the mean difference was usually much less than this and in many subjects the difference between day and night values was small (827, 915). In several of these series the subjects remained recumbent and in one of them (841) they were fasting in addition, but no thorough attempt was made to exclude nychthemeral influences. A less reliable indication of changes in glomerular filtration is given by creatinine excretion, which shows a rhythmic variation very similar to that of inulin (193, 205, 394, 841). In subjects living in isolation in caves and therefore sleeping at unusual clock hours, (607, and unpublished observations upon another subject) the low creatinine excretion rate was observed during the actual time of sleep. It is not, however, solely dependent upon sleep since in subjects living on a 12-hour schedule for a short time (610) creatinine often followed a 24-hour rather than a 12-hour rhythm.

MISCELLANEOUS SOLUTES

There are many reports that total nitrogen (893) or urea excretion is lower by night than during the day (149, 254, 822), but the effect of meals has not always been excluded and there seems to be little evidence for any endogenous rhythm. Low glomerular filtration rate may result in low urea clearance, as may low urine flow, and these two conditions, together with abstinence from food, may well account for the low urea excretion at night which, in the authors' experience, is usually not very far below the diurnal value.

Uric acid is a urinary constituent whose excretion has been less extensively studied within the context of circadian rhythm. Like urea it is at least in part an end product of protein metabolism, whose excretion is likely to depend upon the timing, nature and quantity of food ingested. It is said to be excreted in increased amounts during the day (57, 135, 428) but in all these investigations the subjects appear to have been taking normal meals. Within the experience of the authors the difference between day and night excretion in subjects on a normal existence is often quite large and contrasts with the much smaller differences in urea excretion, but we have found no reports of an endogenous rhythm.

Of minor urinary constituents, organic acids (428) and alkaline phosphatase (394) are said to be excreted in larger amounts by day than by night and amino acids in roughly similar quantities (135), but neither of these has been investigated at all extensively. Those who excrete substantial amounts of protein in urine show remarkable variations in both concentration and excretion rate from one sample to another, but these follow no consistent temporal pattern, even between several days upon the same subject (14). Of much wider potential interest is a study in the time of elimination of a drug after its administration at different times of day (753). The same dose of salicylate was given to each of six subjects at four different times of day. It was most rapidly eliminated when given at 19.00 hours, most slowly when given at 07.00 hours. Such studies have an obvious potential importance in pharmacology and in its therapeutic application, even when nothing is known about the cause of the rhythm and when it has only been observed in subjects living a nychthemeral existence. Another study which could open up a wide field of usefulness is upon circadian variations in the tension of dissolved nitrogen in the urine (3). This is believed to represent a circadian rhythm in arterial nitrogen tension, although both methodological errors and errors arising from variations in renal function could vary circadianly, and so impart a spurious rhythmicity. The main cause, however, is said to be a circadian variation in the circulation-perfusion ratio in the lungs, which becomes more uneven at night. The nocturnal change in urinary nitrogen tension was replicated when a subject lay down during the day, so it appears that this rhythm, if it be genuine, is merely a consequence of the postural alterations in the course of the day. Its real interest is that, by a simple study of urine, information may be obtained which otherwise would involve considerable inconvenience and discomfort to the subject.

The persevering searcher through the journals will find many mentions of urinary rhythms which are not here cited. Some are of minor interest, or merely confirm what is well established; and it is still not always realised (317) that an account of urinary concentrations, however sophisticated the technique of measurement, is of limited value if urine flow is not recorded.

CAUSE OF THE RHYTHMS

When we consider the immediate cause of the different renal rhythms we enter very uncertain territory. Changes in plasma concentration are only well established for phosphate, where, as already described, they are probably responsible for the renal rhythm. By contrast, Renbourn, 1947, (755), carefully examined blood chloride levels in a large number of subjects during the day and could find no evidence of a rhythmic variation; and most workers agree that plasma sodium and potassium concentrations show no rhythmic changes adequate to account for the renal behaviour (606, 615, 677, 841, 915). Wesson (915) found a slight rhythm in plasma potassium concentration; in some subjects this rhythm was absent, and the amplitude of the rhythm of the mean was very small, though it is possible that technical difficulties obscured the existence of a small and consistent rhythm. Plasma potassium concentration is not generally held to be an important determinant of renal potassium excretion (63, 606) and there is no doubt that over long periods of a day and upwards large changes in potassium excretion can occur independently of any change in plasma concentration. Over shorter periods of hours, however, potassium excretion may closely parallel changes in plasma concentration (605) so it is possible that Wesson's figures indicate a real contribution to the excretory rhythm. In a recent account, however (839), a rhythmic change in plasma concentration of potassium, sodium and chloride has been claimed, with phase relations appropriate to the excretory rhythms. There is however even less reason to ascribe control over sodium and chloride excretion to changes in their concentrations in the plasma than there is for potassium, so the significance of this claim for a rhythm in plasma concentration is obscure.

An explanation which has frequently been adduced to account for the acidity of urine at night (124) is a bodily acidity resulting from respiratory depression. It has been shown that a massive increase of CO_2 tension can increase acid excretion by anaesthetized dogs (99, 213, 820) but in conscious man any such effect is small or absent (41, 547). Again it is known that alveolar carbon dioxide tension may rise substantially during nocturnal sleep, but during 24 hours of wakefulness changes in alveolar carbon dioxide tension are trifling, although the oscillation of urinary acidity is as great as when the subject sleeps (604). Moreover, completely paralysed patients whose respiration is maintained by mechanical respirators have perfectly normal excretory rhythms for many weeks (569), despite the constancy of ventilation by day and night. While respiratory depression might therefore contribute to the acidity of night urine, by itself it is quite inadequate to account for it.

Another theory with a certain plausibility is that each phase of the renal excretory rhythms induces a bodily change which results in its reversal. For example, if large amounts of acid are being excreted at night the body fluids would be left somewhat to the alkaline side of neutrality and adaptive changes would then cause the excretion of an alkaline urine which, if there were sufficient delay between the stimulus and the response, might overshoot and result in the excretion of an excess of alkali and a body somewhat on the acid side, causing reversion of the urine to the acid state. Such a system, known technically as relaxation oscillation, tends to have a fairly fixed amplitude but variable period, which is just the reverse of the usual renal excretory rhythms. The theory is disproved by experiments (622) in which large doses of alkaline potassium salts were taken on three consecutive evenings, so that by night the kidney excreted an alkaline urine rich in potassium. So soon as the subjects ceased this dosing, the renal excretory rhythms for potassium and for acid immediately reverted to their normal phase and pattern although for the previous three days the usual oscillations had been absent. It seems plain therefore that the kidney was controlled by some rhythmic influence upon it. The possible rhythmic influences must therefore be considered in turn.

The possibility of a nervous influence is largely unexplored, and the nervous system is commonly neglected as a potent influence upon renal behaviour. It is claimed that normal rhythmicity persists in a homo-transplanted kidney (314); but observations upon two further patients, with transplants from cadavers (64), have shown a reversal of the usual rhythm; GFR, flow and electrolyte excretion were high at night and lowest in the period 0800-1400 hours. A preliminary note from another group (9) confirms this finding upon another nine patients, including one whose transplant was from a monozygotic twin and who was receiving no steroids nor other drugs. The habits of the patients are not mentioned, and these reversed rhythms may be wholly exogenous. Phosphate excretion, however, was higher by night, as in normal subjects. This is of particular interest since the phosphate excretory rhythm seems to be controlled extrarenally by the plasma concentration.

Of chemical influences, we should consider mainly the plasma concentrations of substances excreted, and hormones which act upon the kidney. We have seen already that of the solutes whose excretion varies rhythmically, only phosphate and perhaps potassium show rhythmic changes in plasma concentration. Plasma concentrations of sodium and chloride, by contrast, remain remarkably constant throughout the 24 hours. Since we are largely ignorant of the factors which control renal excretion of these electrolytes even in ordinary homoeostatic situations, there is little hope as yet of an adequate explanation of the kidneys' rhythmic behaviour. The influences upon the kidney which are known to be rhythmic are glomerular filtration rate and the secretory activity of the adrenal cortex. Glomerular filtration rate is believed by some (680) to be a major factor determining the rate of sodium and chloride excretion, whilst others (525) consider it to be at most of minor importance. The circadian variations in glomerular filtration rate considered above (Page 59) have been sometimes of sizeable magnitude, sometimes trivial, and large regular variations in sodium excretion have been observed equally in subjects whose glomerular filtration behaves in either way. Indeed when individual subjects are considered (882) the correlation between glomerular filtration rate and sodium excretion is poor and a similar lack of parallelism between excretion of sodium and creatinine (607) points in the same direction.

ADRENAL CORTEX

The part played by the adrenal cortex is far from clear. Normally the secretion of cortisol, at least as assessed by its plasma concentration, reaches a peak in the early morning and its excretory rate a similar peak about three hours later (see Chapter 3). Cortisol is known to increase potassium excretion (48, 246, 618, 620) and in small concentration may also increase sodium excretion (419), and the time relations of the morning increase are about right. Observations on the persistence of excretory rhythms in the absence of the adrenal cortex (193, 273, 511, 521) or in the presence of continuous high cortical activity, either spontaneous (205) or due to continuous administration of large amounts (775), are often cited as an indication that the adrenal cortex is not involved in these rhythms. Many of these observations are inconclusive, in that potential rhythmic influences such as posture were not controlled, but in some (273, 521) the subjects remained recumbent, with uniformly spaced meals, and showed an essentially normal electrolyte excretory rhythm in the absence of rhythmic adrenal cortical secretion. It remains possible that the alternation of wakefulness and sleepiness induced by light and darkness and social contact had exerted a rhythmic influence upon the kidney, by some unknown intermediation, and independently of a central clock; but, even if they are not absolutely conclusive, these observations weaken the evidence that the adrenals play an essential role in transmission between the clock and the kidneys.

Inhibition of the adrenal excretory rhythm with drugs prevents the morning rise in excretion of sodium and potassium (419). This has been achieved by continuous bombardment of the hypothalamus and pituitary with potent steroids, and by blocking the 11-β-hydroxylase enzyme involved in the synthesis of steroids. It is probable that both these procedures block the production of cortisol rather than of aldosterone. It has been suggested that this merely indicates a 'permissive' role of the adrenals, that in the absence of their secretion the rhythmic influence, whatever it may be, is unable to exert its effect; but this explanation would hardly hold for the disappearance of rhythmicity under constant steroid influence. When different renal rhythms become dissociated the excretion of potassium often remains associated with that of adrenal steroids. Thus these two components are the slowest to adapt to a change of phase in external rhythms, achieved by flying into a different time zone, or by sleeping by day (251, 570), though we have observed the potassium rhythm to adapt much more rapidly than that of the adrenals after transatlantic flight (164, 229). When night workers whose renal rhythms have or have not adapted themselves to the change of routine are compared, we have found good adaptation of the kidney associated with adaptation of the adrenal and also dissociation between the rhythms of excretion of potassium and of 17-hydroxycorticosteroids as has been claimed in subjects living on days of abnormal length (826). The subjects spent some weeks in the Arctic living on a 'day' of 21 hours, and the presence or failure of adaptation of the different rhythms was assessed by their conformity to a rhythm of 21 or 24 hours. 'Adaptation' could represent the response to any one of a number of influences following a 21-hour cycle, but the supposed internal clock was the only known influence with a 24-hour period. Thus, the necessary evidence for the unimportance of an adrenal rhythm as a cause of the potassium excretory rhythm would be a clear persistence of 24-hour periodicity in potassium excretion in association with a loss of such rhythm for 17-hydroxycorticosteroid excretion. Such dissociation cannot be found in the published data. The dissociation observed on several occasions (825) was between a 21-hour rhythm for potassium excretion and a 24-hour rhythm for 17-OHCS excretion; potassium excretion was thus not determined by the adrenals, but it may have been determined by a 21-hour rhythm in electrolyte intake, or by some other aspect of the subject's 21-hour routine.

Another line of argument also casts doubt upon the adequacy of the adrenal rhythm to explain the potassium excretory rhythm. If the matutinal excretion of potassium is due to the rhythm in cortisol secretion, it should be possible by administration of cortisol around the time of lowest excretion, midnight or so, to raise it to the level normally found in the morning. Attempts to do so with doses of cortisol comparable with those believed to be liberated by the adrenal cortex (418) have, however, been only partially successful.

Aldosterone is unlikely to be an important factor in excretory rhythms of sodium and potassium. Although its urinary excretion (651, 652, 653, 930), plasma concentration (931), and its rate of secretion by the adrenals (915) have been shown to vary circadianly, the secretion rate is greater in the morning when sodium excretion is high, the opposite to that which would be expected if aldosterone influenced the sodium excretory rhythm. Moreover the aldosterone rhythm seems to be almost entirely dependent upon the postural changes associated with normal activity whereas the electrolyte excretory rhythms continue independently of such postural changes.

INFLUENCES ON SODIUM EXCRETION

It has become increasingly recognized that a number of factors hitherto undiscovered have important effects on sodium excretion (168, 434, 742), and that these include some humoral agent which is not of adrenal origin (177, 194, 429, 519). Among those to be identified are growth hormone, which causes sodium retention and which also modifies the response of the kidney to aldosterone (544, 545), angiotensin which has a direct effect upon the renal tubules (542) as well as indirect effects by stimulating the secretion of aldosterone, a steroid liberated from the heart when venous return diminishes, a non-steroid hormone liberated from the lungs (417, 543) and bradykinin (44, 287). Vascular changes within the kidney can also influence sodium excretion by some means other than change of GFR (181, 600). Little or nothing is known of any possible circadian rhythmicity in these factors. In the absence of such knowledge speculation on the immediate circadian influences operating upon the kidney would be premature, but it would appear that although cortisol exerts an important influence, it operates upon a different background, perhaps a different endocrine background, in the morning from its field of operation at night.

The gradual fade-out of the previous circadian rhythmicity in subjects exposed to a new rhythm of habit is commonly referred to as entrainment. The supposition is that the new habit is increasingly exerting its influence either upon the kidney itself or upon the controlling mechanisms somewhere further back in the causal sequence, so that ultimately the influence of the previous circadian clock has disappeared. Another kind of 'fade out' is rather more puzzling. It has been mentioned already that the usual excretory rhythms persist in completely paralysed subjects whose ventilation is mechanically controlled at a steady level throughout the 24 hours. However, after two to four weeks of such artificial ventilation the potassium excretory rhythm gradually disappears, and the sodium rhythm vanishes some time later (569). A similar fade-out has been observed in subjects given steady dosage of cortisone (775) for some days. Two possible explanations may be advanced: either the rhythmicity of the 'clock' is fading out, or its influence upon the kidney is waning. Observations on subjects deprived of all external Zeitgeber, discussed in Chapter 7, indicate that reinforcement from external Zeitgeber is necessary if the clock is to keep accurate 24-hour time. There could be an autogenous feedback, whereby the clock, in imparting rhythm to the adrenals and to pulmonary ventilation, applies to itself a 24-hour rhythm in the composition of the blood bathing it; if this feedback failed, and the clock stopped, one would expect not only the renal rhythms but all spontaneous rhythms to disappear. The alternative explanation would be to suppose that the influence of the clock upon the kidney needs constant slight reinforcement by other circadian influences, of which variation in pulmonary ventilation or in adrenal secretion may be important.

Here, however, we are penetrating unduly the realms of speculation. We are unlikely to understand the sodium excretory rhythm until we have a much clearer understanding of the regulation of sodium excretion in other circumstances.

(A convenient tabular bibliography of renal circadian rhythms in man has been published, 915.)

Chapter 5

Cardiovascular, Haemopoietic and Respiratory Rhythms

Cardiovascular: Blood pressure; Peripheral Vascular Resistance and Blood Flow; Cardiac Output; Heart Rate; Performance-Pulse Index and Exercise—Haemopoietic: Leucocytes; Erythrocytes; Haemoglobin; Other Blood Proteins and Specific Gravity; Blood Clotting Mechanisms —Respiration: Ventilation; Vital Capacity, and Airway Resistance; Diffusing Capacity; Origin of Respiratory Rhythms.

We gather together in this chapter cardiovascular, haemopoietic and respiratory circadian rhythms. Most of these appear to be exogenous.

CARDIOVASCULAR SYSTEM

Circadian periodicity has been noted in many circulatory functions including blood pressure, pulse rate, plasma volume and cardiac output (437, 586). It has been suggested, indeed, that all circulatory values follow a similar 24-hour rhythm (488, 489). Certainly they are closely related to both one another and also to other circadian rhythms and it is difficult to determine the precise extent to which they are exogenous or endogenous. Many of them have been found, however, to be independent of the sleep/wakefulness cycle and to persist in subjects kept recumbent and restricted to small regular meals (488, 489, 581). Rhythms of blood pressure and heart rate have been shown to persist in a subject in coma (508) but, whereas the heart rate remained synchronized to 24-hour time, the blood pressure was desynchronized and showed a 24·6-hour period.

BLOOD PRESSURE

Howell in 1897 (405), using a plethysmograph, reported a fall in blood pressure during sleep. It was found that a marked fall in blood pressure during the first hour of sleep may be followed by a gradual rise during the night and an abrupt increase after waking (126). A fall in blood pressure when night workers slept during the day has also been noted (108). During the day blood pressure rises gradually in both hypertensive and 'control' subjects (649).

The existence of a rhythmic variation in blood pressure has been questioned. Only transient post-prandial increases have been reported in ambulant subjects (923) and no fall at night was seen in subjects receiving continuous intravenous infusions and 4-hourly

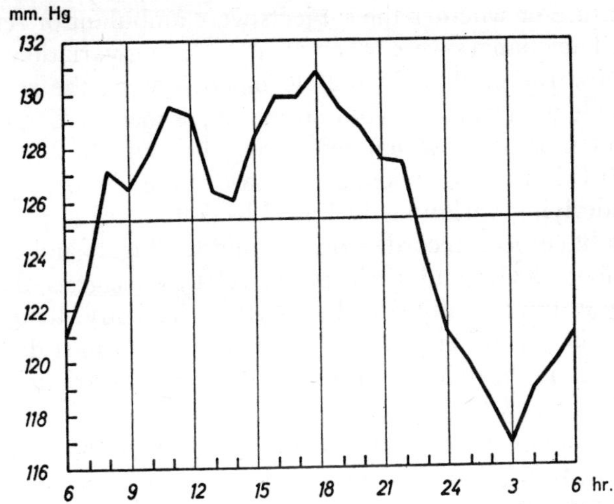

FIG. 5.1. Variations in systolic pressure over the 24 hours. A summation curve from 120 patients with a normal circulation. (Zülch, K. H. and Hossmann, V., Fig. 2A, 945.)

venepunctures (106). Records with an automatic oscillograph have demonstrated peak values in the early evening and a fall, sometimes marked, during sleep (759).

A circadian rhythm of more complex form was seen in 120 normotensive subjects (Fig. 5.1). In addition to a deep trough about 03.00 hours, a smaller trough was seen at 14.00 hours. Peak values occurred around 12.00 hours and 18.00 hours. A similar curve was found in a nineteen-year-old subject who had been decerebrate for seven years as a result of an eight-minute cardiac arrest, which suggests that there is a control system for the blood pressure rhythm at the hypothalamic or brain stem level (945).

Continuous blood-pressure recordings from an intra-arterial catheter for up to 23 hours (67) have shown much larger random variations than had hitherto been suspected, with a slight fall in the course of the day and a large fall during sleep.

PERIPHERAL VASCULAR RESISTANCE AND BLOOD FLOW

Rhythmic variations in limb skin temperature have been reported both in normal subjects and in subjects recovering from hypothermia (80). Regular circadian changes were seen with minimum values between 08.00 and 16.00 hours and maximum values between 20.00 and 04.00 hours irrespective of ambient temperature or whether the subjects were ambulant or remained in bed. The changes were taken as representing variations in skin blood flow rather than in muscle blood flow as the recordings were made at sites away from large muscle masses. Large differences in temperature were noted during the course of the 24 hours, up to 10°C in readings taken at the base of the big toe, suggesting that underlying variations in limb blood flow are large. Similar changes have been recorded on the middle finger, and the back of the foot (386), and their phase relation suggests that they contribute importantly to the rhythm in body temperature rhythm, the rise of temperature being at least in part due to the diminution in heat loss resulting from vaso-constriction in the extremities (see Chapter 2).

Measurement of forearm blood flow by venous occlusion plethysmography has also indicated rhythmic changes in blood flow. In a group of eight subjects in whom resting and maximum post-exercise blood flow was noted at 09.00, 11.30, 15.00 and 22.30 hours, peak values in both were seen in the late afternoon or early evening and a cyclic pattern in peripheral resistance indicated (451) which is out of phase with skin flow.

Capillary resistance as measured by petechial counts has been reported to show a 24-hour rhythm with a trough about 23.00 and a peak at 05.00 hours (211, 752).

CARDIAC OUTPUT

In view of the rhythmic changes in blood pressure and peripheral resistance it is not surprising that a circadian variation in cardiac output occurs. Maxima in stroke and minute volume have been reported between 12.00 and 16.00 hours and minima between 02.00 and 04.00 hours (449). A fall from an output of six to seven litres a minute during the day to four litres per minute at night has been noted (106).

HEART RATE

Detailed studies by Kleitman and Ramsaroop (476) of a group of six subjects examined at 2-hour intervals, have shown that the heart rate follows a circadian rhythm with lowest values at night. Shifting of times of sleep and activity have been found to be quickly accompanied by adaptation of the heart rate.

The same workers also present extensive data indicating that the heart rate and body temperature curves can be closely superposed except when a post-prandial rise in pulse rate occurs. A rise of 10 to 15 per minute in the pulse rate accompanied a 1°F increase in rectal temperature (476). Since the body temperature can affect the heart rate directly, the change in heart rate may have been a direct consequence of the change in body temperature. Parallel variations in pulse rate and body temperature which they noted after administration of thyroid hormone could well be due to two independent effects of the hormone since the rise in pulse rate per minute was 26 per degree increase in temperature under these circumstances.

Dissociation between the two rhythms has been shown in subjects living on abnormal time schedules in the Arctic (475). Kleitman's own temperature showed a persistent 24-hour rhythm when he was living on an 18- or 28-hour day whereas his pulse rate quickly entrained to either of these schedules. There were, however, many environmental cues which may have helped to maintain the body temperature rhythm.

The rapid adaptation of the pulse rate to abnormal time schedules or shifts in times of sleep and activity suggest that it is not an endogenous rhythm but is exogenous, governed mainly by habit, increasing during activity and decreasing during rest and sleep. In the absence of other major influences, as in subjects kept under fairly constant conditions, pulse rate may be largely determined by body temperature, and thus appear to show an endogenous rhythm. This may account for the fall in pulse rate recorded in pilots during flights at times when they would normally be asleep (406), but unfortunately their temperature was not measured.

In pregnancy the rhythm is maintained with maximum values at 09.00 hours and minima about 03.00 hours (561, 565).

PERFORMANCE-PULSE INDEX AND EXERCISE

The performance-pulse index (the ratio between increase in pulse rate and increase in work output) has been shown to vary in a circadian manner (891). Minimum values between 02.00 and 04.00 hours and a peak between 16.00 and 18.00 hours were found in a group of 20 subjects using a bicycle ergometer. This suggests that there is a circadian influence which summates algebraically with the influence of exercise. Similarly measurements of cardiac frequency, ventilation rate and temperature in

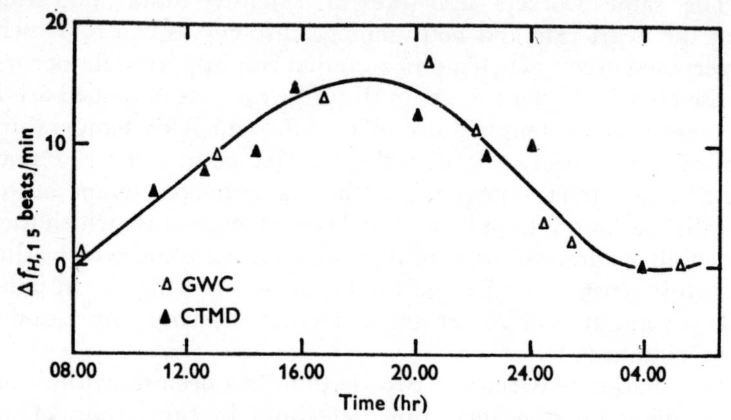

FIG. 5.2 Circadian variation in cardiac frequency in response to sub-maximal exercise on a bicycle ergometer (Crockford, G. W. and Davies, C. T. M., Fig. 1, 171.)

two subjects submitted to 20-minute continuous exercise tests, on a bicycle ergometer, every 90 minutes for 27 hours, also showed a circadian rhythm with minima about 05.00 hours and maxima in the afternoon in all three variables (Fig. 5.2).

HAEMOPOIETIC SYSTEM

Circadian rhythms occur in the haemopoietic system; the eosinophil rhythm is the largest and best defined, but other leucocytes, the erythrocytes, haemoglobin and the blood clotting mechanisms also show circadian periodicities.

LEUCOCYTES

Japha in 1900 (425) noted that daily variations in total numbers of leucocytes took place independently of food and that high levels were reached in the afternoon followed by an evening fall. The occurrence of a rhythmic variation in the leucocyte count was confirmed by Sabin and her colleagues (783), who also noted peak values in the afternoon. A nocturnal peak may occur after midnight, and the rhythm is independent not only of food but also of exercise and sleep and is not correlated with the body temperature rhythm (814).

Von Domarus, 1939 (894), and Djavid, 1935 (201), observed high morning values in eosinophil levels; Appel, 1938 (17), noted minimum values in these cells at 10.00 hours and an evening rise

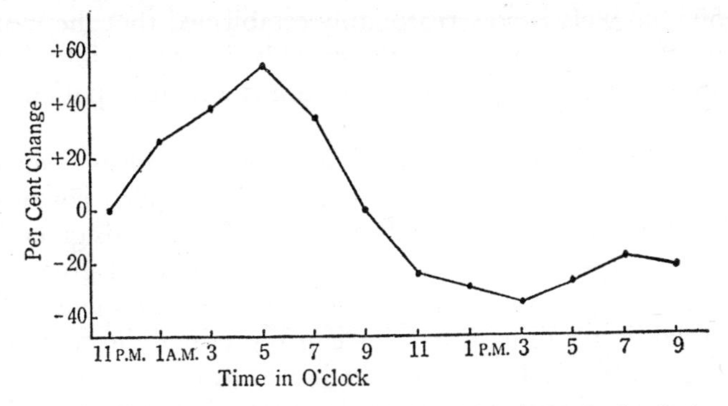

FIG. 5.3. Circadian variation in average level of circulating eosinophils. (Tatai, K. and Ogawa, S. Fig. 3, 857.)

but his observations were few in number. He also found, however, that the morning rise occurred earlier the earlier the sun rose. Rud, 1947 (779), in the first study extending throughout the 24 hours, made counts every 15 minutes during the day and hourly at night on a group of four subjects. He found that a decline to a minimum at 09.30 hours was followed by a steady rise till about 13.00 hours and then by fluctuating but gradually increasing levels throughout the day with highest levels between 24.00 and 03.00 hours. These findings have been confirmed and elaborated by a number of investigators (18, 88, 105, 210, 249, 381, 392, 456, 561, 565, 852, 857).

The common practice of expressing changes in eosinophil count as percentage (Fig. 5.3) rather than numerical values has been justified by a careful statistical analysis (4), and the assertion (567) that different individuals have their characteristic patterns was also validated. Many of the earlier workers confined their investigations to a normal working day, whereas the fall between 06.30 and 09.30 (332, 334, 349) may be more striking and more consistent than the changes later in the day. However, the changes in eosinophil count during the 24 hours can be reasonably and significantly described by a sine curve (347).

Elmadjian and Pincus in 1946 (230) noted the close temporal relationship between changes in eosinophil levels and the excretion of 17-ketosteroids, and the association of the eosinophil rhythm with changes in adrenal cortical function has been studied in considerable detail, in both health and disease, particularly by Halberg and his colleagues (249, 252, 320, 321, 322, 336, 347, 349, 595, 889). It is now reasonably established that the morning rise in the secretion of corticosteroids is responsible for the morning fall in eosinophil levels, which is absent in patients with Addison's disease.

The action of the adrenals is not merely permissive, since the eosinophil rhythm is not restored in patients given cortisone in divided doses throughout the 24 hours (446). Steroids given in the morning (387, 668) cause a fall in eosinophil count greater than that which other workers have observed to occur spontaneously, and in Addison's disease steroids can reproduce the spontaneous morning fall (446). There appears, however, to be no account of a comparable eosinopenia induced by adrenal steroids at a time of day when it does not usually occur.

The eosinophil rhythm also appears to be highly light-dependent. The manner in which the morning fall follows the season of the year was remarked on by Appel in 1938 (17) and it was later shown (18, 499) that in blind subjects the morning eosinopenia may be delayed or diminished. Blindfolding produced a similar effect (509, 809). It has been reported that night workers do not show reversed eosinophil rhythms (17, 598). However, subjects in the Arctic who reversed their sleep-activity pattern, and who slept in a blacked-out tent and thus received an intense light stimulus on emerging, showed reversed eosinophil rhythms in less than three days (808). Radnot and his colleagues (734, 735, 736, 737, 738), have also made a number of investigations into the effects of light on the eosinophil cycle. They report that subjects kept in darkness showed no morning eosinopenia, and that the admission of light to the eyes was followed promptly by eosinopenia either in the morning or in the afternoon. Since adrenal activity is not so closely dependent upon light, they do not consider that cortisol can mediate the light-induced eosinopenia, and conclude that the morning fall in eosinophil levels is directly due to illumination.

An inverse relationship between body temperature and eosinophil levels has been noted (339, 392).

A circadian rhythm in circulating lymphocytes also occurs. A sharp fall on waking is followed by a steady increase throughout the rest of the 24 hours (230). In pregnancy a similar change has been reported (561, 565). Administration of steroids can induce lymphopenia but the change is smaller and less regular than with eosinophils (668).

The rise in total leucocyte count and fall in eosinophils and lymphocytes during the course of labour is said (561, 565) to represent an exaggeration of the usual circadian rhythm in these cells, since labour usually begins around 03.00 hours (see Chapter 8). It would be interesting to compare these changes with those when labour starts at a different hour.

A circadian variation in neutrophils characterized by a steady increase throughout the day, and a morning fall and evening rise in basophil levels, have been reported (561, 565, 808). Reversal of sleep and activity is followed by reversal of lymphocyte and basophil periodicities within three days as with the eosinophils. The neutrophil pattern takes rather longer to change (808, 812). The effects of alterations in the light routine on leucocyte variations have also been studied by Sharp (809) who has shown that by keeping individuals in darkness for three hours after waking, the normal morning changes in leucocyte counts are postponed (Fig. 5.4). It is claimed that the timing of the leucocyte rhythms is modified in pregnancy (561, 565).

ERYTHROCYTES

24-hour variations in erythrocyte levels have been recorded (98, 504, 907). Similar changes in the reticulocyte levels and erythropoiesis have also been suggested (292, 294) but can hardly contribute significantly to the variations in erythrocyte count. Price-Jones in 1920 (726), described daily variations in the sizes of red blood cells. Variations during the 24 hours in erythrocyte sedimentation rate have also been noted but do not appear to be circadian (437, 755). A circadian variation in the haemocrit levels (581) and in blood viscosity (805) has also been reported.

HAEMOGLOBIN

That a fall in haemoglobin levels occurs between morning and evening has been known for many years. Ward, as long ago as 1904 (907), noted variations from hour to hour in the quantity of haemoglobin in human blood. The variation is a 24-hour one and occurs irrespective of whether a subject stays in bed or has

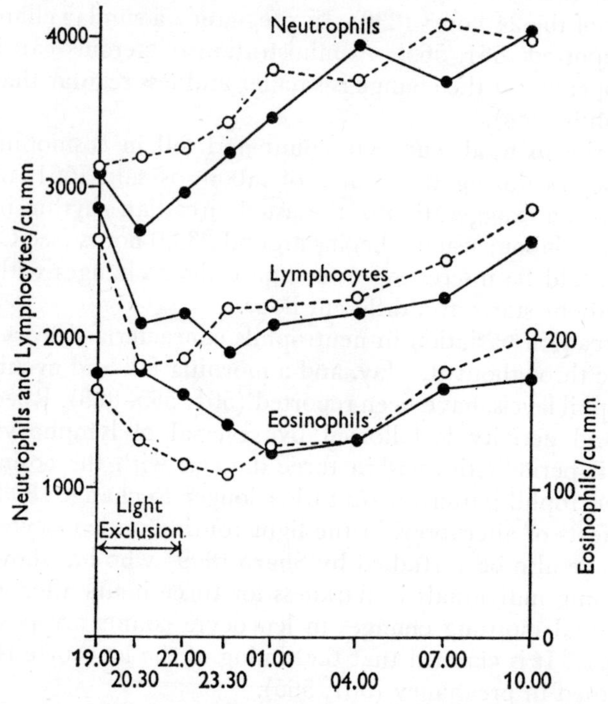

Fig. 5.4. Variations in leucocyte counts under control and light exclusion conditions. (Sharp, G. W. C., Fig. 1, 809.)

no rest for over 30 hours or is pyrexial (217). A drop between 08.00 and 18.00 hours has been claimed (731). In another investigation (657) samples collected at 09.00, 11.00, 14.00, 17.00, 20.00 and 23.00 hours showed a small but continuous fall. Variance analysis on these figures showed that up to 17.00 hours the fall was not significant, but it had become significant by 23.00 hours (630).

This variation in haemoglobin occurs irrespective of whether a subject remains in bed until evening and then rises and works or spends the day working and goes to bed at an early hour (755). The magnitude of the change can be quite large. A decrease of 0·34 g. Hb/100 ml blood between 09.00 and 15.00 hours has been noted (906). Ten subjects studied over five successive days showed an average decline of 0·3 g. Hb/100 ml from samples taken between 09.00 and 10.00 hours to those obtained between 16.00 and 17.00 hours (232).

OTHER BLOOD PROTEINS

Total protein-bound carbohydrates, hexosamine and sialic acid all show circadian variations with minimum values between midnight and 04.00 hours (805).

SPECIFIC GRAVITY

Jones in 1889 (432), demonstrated a 24-hour rhythm in blood specific gravity with highest levels occurring between 09.00 and 10.00 hours and lowest from 18.00 and 19.00 hours (432, 433). The average daily variation has been stated to be 0.0033 in men and 0.0027 in women (504) and can be reasonably considered as largely due to variations in the numbers of red blood cells.

Changes in the concentration of any blood constituents may result from changes in addition or removal or may be a passive consequence of changes in blood volume. Haemoglobin, and the erythrocytes which contain it, have a life of some 3 or 4 months, so that any changes due to addition or removal could only involve temporary sequestration in organs such as the spleen. It is more likely that circadian variations in haemoglobin are due to variations in blood volume, which have been directly recorded and show the appropriate phase relationship (169, 247). For constituents with a short life, such as leucocytes, the rhythm is more likely to reflect changes in production or removal; and this is especially true where the circadian oscillation is large, as with eosinophils.

BLOOD CLOTTING MECHANISMS

A rise from 10.00 hours to a maximum at 16.00 to 18.00 hours in platelet levels has been noted in a group of five subjects examined at 2-hourly intervals from 10.00 to 22.00 hours (483). In studies in which blood samples have been obtained at 3-hourly intervals throughout the 24 hours the rhythm has been shown to be circadian with lowest values occurring during the night (293).

A daily variation in platelet stickiness has also been noted (66). Twenty-two normal subjects and 40 patients with ischaemic heart diseases were studied. Measurements were made two to five times during the day. A significant rise in stickiness from 11.00 to 13.00 hours and again to 15.00 hours was seen in 80 per cent of the subjects. Food, age, sex, menstruation and physical activity had no apparent influence but a possible relationship to catecholamine output was suggested.

A daily variation in plasma fibrinolytic activity has been recorded by a number of observers (68, 128, 242, 319, 568). A marked nocturnal fall, as judged by lysis times, was found in two subjects from whom blood samples were obtained at intervals throughout the 24 hours (242). In a comparison between nurses on day and night duty, twelve of fifteen day nurses had a significantly longer lysis time, i.e. less fibrinolytic activity, at 04.00 than at 16.00 hours; 14 out of 15 night nurses behaved in the same way.

Fibrinolysis times have been measured in 53 subjects at 09.30 and 11.30 hours after a fat-free breakfast (68). Twenty-five of the subjects were confined to bed, the other 28 were ambulant but rested for half an hour before the test. Lysis times (expressed in hours) in the active group showed a mean value of $3{\cdot}75 \pm 0{\cdot}28$ at 09.30 hours and $2{\cdot}93 \pm 0{\cdot}17$ at 11.30 hours, a change of $0{\cdot}82 \pm 19$ which was significant ($p < 0{\cdot}001$). Lysis times in the group who were confined to bed were $3{\cdot}34 \pm 0{\cdot}28$ and $3{\cdot}06 \pm 0{\cdot}22$, at 09.30 and 11.30 hours respectively, a non-significant difference of $0{\cdot}28 \pm 0{\cdot}17$. In another study, 80 subjects were tested at 10.00 and 16.00 hours, and 16 were further examined at these times on five consecutive days. Mean blood clot lysis time fell between 10.00 and 16.00 hours but mean plasma anti-urokinase activity was not affected by time of day (568).

It would appear that the increase in fibrinolysis in the course of the day is exogenous, since it is absent in recumbent subjects; but the observations upon night nurses suggest that the prolonged times in the night are in part endogenous. There is obvious need for more records at intervals spaced throughout the 24 hours, since the above conclusions are each based upon a single series.

RESPIRATION

A number of respiratory circadian rhythms have been described; all of them appear to be exogenous in origin.

VENTILATION

During sleep characteristic alterations in breathing are known to occur. They consist mainly of a fall in ventilation with a consequent rise in alveolar carbon dioxide tension (130, 233, 234, 382, 383, 384, 437, 733, 744, 846, 875). Respiratory minute volume decreases by about two litres a minute during sleep and the CO_2 sensitivity curve is shifted to the right, both at sea level and at altitude (743). These changes may be largely ascribable to sleep, but as one of us has demonstrated (604), a small rise in alveolar carbon dioxide tension also occurs during sleepless nights. This suggests a circadian rhythm in the central drive to the respiratory centre (130, 557, 689, 771).

The difference in tension between alveolar and urinary nitrogen is believed to reflect inequalities in the ventilation-perfusion ratio in the lungs. In a subject examined over the course of several weeks the difference was significantly lower during the day than at night. Recumbency, however, was found to increase the urinary/alveolar nitrogen difference and thus could be responsible (3).

VITAL CAPACITY, AND AIRWAY RESISTANCE

Vital capacity falls during the day to a minimal value around midnight, and then climbs until about the usual time of waking (486, 487, 581). This rhythm persists in subjects continuously recumbent, and taking meals every 4 hours, and is not dependent upon sleep. The immediate cause of the changes is a matter for speculation, but the alertness and adrenal medullary rhythms have been suggested.

As a measure of airway resistance, the one-second forced expiratory volume (FEV_1) has been measured at frequent intervals in a small group of normal subjects (510) and at 8-hourly intervals in a much larger group (903). Minimum values in both groups were found at 06.00 hours.

DIFFUSING CAPACITY

The diffusing capacity of the lungs (transfer factor) also exhibits a daily variation (143). In 24 normal subjects in whom the pulmonary diffusing capacity for carbon monoxide (DL_{CO}) was measured at 2-hour intervals from 09.30 hours to 21.30 hours, a continuous fall was observed; it was also seen in subjects who remained lying down from 09.00-16.00 hours. The authors considered that not more than 40 per cent of the reduction in DL_{CO}

was due to circadian variations in haematocrit or haemoglobin levels and that variations in catecholamine release might also be involved since a rise in DL_{CO} has been observed after noradrenaline infusion (512), and catecholamines show a nychthemeral rhythm (Chapter 3). It remains to be investigated whether this observed rhythm is wholly exogenous.

ORIGIN OF THE RESPIRATORY RHYTHMS

A catecholamine rhythm could account for the variations in airway resistance and diffusing capacity. For many other respiratory rhythms a likely cause is the rhythm in alertness (Chapter 7). During a high degree of alertness, respiratory changes opposite to those which are seen in sleep have been demonstrated (130), suggesting the importance of a cerebral influence on the respiratory centre (602). Thus circadian variations in respiratory function may be closely linked to the sleep/wakefulness rhythm. This appears to be confirmed by experiments in which respiratory function was found to follow closely an abnormal pattern of sleep and wakefulness.

Two subjects were confined for nine days in a chamber in which temperature was maintained at 27°C ±0·1°, humidity 30 per cent ± 5 per cent, and barometric pressure at 30·56 ±004 inches. Respiratory frequency was monitored continuously by telemetry. Lung volumes, vital capacity, inspiratory capacity, and expiratory flow rate were measured four times a day. The subjects shifted 1·75 hours per day away from clock time during the nine days. The various lung functions measured shifted with the sleep/wakefulness cycles except for the respiratory rate which became dissociated in both subjects and tended to show a predominance of 6 hour frequencies (146, 147, 790, 791).

It is thus doubtful whether any respiratory variable has yet been shown to have a clearly endogenous and independent rhythm. All these variables may, indeed, depend on the alertness rhythm, acting both directly upon the respiratory centre and indirectly through the adrenal medulla.

Chapter 6

Miscellaneous Rhythms

Metabolic: Carbohydrates, Glycogen, Fats, Alcohol, Amino-Acids—Serum Enzymes—Electrolytes: Iron, Copper—Digestive Tract: Saliva, Dental Tissues, Gastric, Biliary, Pancreatic, Colonic—Cellular: Mitosis—The Eye: Intraocular Pressure, Pupillary Size.

A wide variety of metabolic and cellular rhythms have for convenience been brought together in this chapter. For very few of these is there any evidence that they are endogenous so it is very much a descriptive chapter.

METABOLIC RHYTHMS

A circadian variation in metabolic rate in man has been described (89, 488, 489, 893) which persists in subjects kept fasting and at rest with minimum values in the early morning. This rhythm may well be a result of the circadian rhythm in body temperature.

CARBOHYDRATE RHYTHMS

These have been extensively investigated by Möllerström (631, 632, 633, 636, 638, 639). A daily variation in blood and urinary sugar levels in both diabetic and normal subjects has been described by him with low levels occurring about noon, despite the intake of food. He has recorded a rhythmical variation in β-hydroxybutyric acid excretion in diabetics, and a nocturnal fall in citric acid output.

A greater post-prandial rise in blood sugar levels after the midday meal than after breakfast has been noted (305, 925) and in a detailed study of 57 patients in whom both morning and afternoon testing was carried out, a change in glucose tolerance between these two times has been found (768). The occurrence of a daily variation has been confirmed by Bowen and Reeves (92) who examined twenty normal subjects, sixteen females and four males. Blood samples were taken before giving 75 grams of glucose to each subject. Further blood samples were then obtained at hourly intervals for 3 hours. The test was performed on all subjects at both 08.00 hours and 12.00 hours. The 1-hour and 2-hour blood sugar levels after taking glucose were consistently higher in the afternoon. These blood sugar changes may be explicable in terms of a rhythmic variation in insulin levels, which, as has been described in Chapter 3, are higher in the morning and fall in the afternoon.

GLYCOGEN

Some liver biopsy studies have been reported (63) as showing higher glycogen levels in the afternoon than in the morning, but other observers (85) have found no difference at these times.

FATS

Cholesterol

Daily variation in total serum cholesterol has been investigated by a number of observers. Some have noted little change (94, 123, 878) but others (139, 659, 709) have reported very marked hour to hour variations. Studies by Page and Moinuddin (690) and by Hollister and Wright (398), however, indicate that serum cholesterol levels are remarkably stable during the day and the evidence would appear to be very much against the occurrence of a circadian variation.

Lipids

A circadian rhythm in the urinary excretion of neutral reducing lipids was noted by Pincus *et al.* (719) with lowest values at night and maximal excretion in the morning.

Daily changes in serum triglyceride concentrations have been described (490). These were seen in both lipaemic and hyperlipaemic subjects on various test meals. Circadian rhythms in blood lipids and ketone bodies in young diabetics on fat-rich diets have also been noted (845) and a 24-hour rhythm in chylomicrons has been reported (86). In a study of 13 diabetic patients and one non-diabetic subject, Maruhama *et al.* (574) obtained blood samples at 2-hourly intervals for 24 hours from 08.00 hours onwards. An 1800 calorie test meal divided into three parts was given at 08.00 hours, 12.00 hours and 17.00 hours. In both the healthy and the diabetic subjects a circadian variation in blood sugar and plasma triglycerides was seen.

Serum phospholipid levels have been found to be quite stable during the day (398).

ALCOHOL

A daily variation in alcohol metabolism has been reported (927). Six subjects were given small doses of ethanol, diluted with water to a final alcohol concentration of 20 per cent by volume, over at least two 36-hour periods. The alcohol was administered hourly except that the 01.00 and 02.00 hours and 04.00 and 05.00 hour doses were combined. Blood or saliva samples were collected before each dose. Alcohol concentrations in these fluids rose from 20.00 hours to 10.00 hours and then fell again until 20.00 hours, except in one subject who was on night duty and in whom the pattern was reversed.

The variations appeared too great to be due to changes in excretion and too prolonged to be accounted for by changes in absorption and were considered to suggest a change in rate of combustion related to the sleep/wakefulness cycle.

AMINO-ACIDS

Feigin and his colleagues have demonstrated the existence of a circadian periodicity in whole blood and serum amino-acids in man, with minimum concentrations at 04.00 hours (Fig. 6.1). Individual amino-acids showed the same periodicity. The wide range of normal values for amino-acids reported in the literature may well, as these authors suggest, be caused in part by differences in sampling time (243, 244).

Daily rhythms in amino-acid concentrations have also been described by Wurtman *et al.* (938, 939). Twenty-three young male subjects aged from 18 to 25 years were investigated in one experiment. They were restricted to one of three diets each having the same number of calories but differing in the amount of protein. One contained 1·5 gm of protein per kilogram body weight, the second 0·71 gm, and the third less than 0·04 gm. A series of seven blood samples were obtained at 3- to 6-hour intervals over the course of twenty-four hours starting at 07.00 hours. All samples were analysed for tyrosine, tryptophane and phenylalanine. Threonine, serine, glutamic acid, glycine, alanine, valine, cysteine, isoleucine, leucine, lysine, histidine, and arginine were estimated in two samples from each subject, the samples chosen being those which had the highest, and lowest, concentrations of tyrosine respectively; methionine concentrations were assayed in blood samples from two subjects who were on the diet lowest in protein.

Tyrosine, tryptophane and phenylalanine concentrations showed a rhythmic variation throughout the day with minimum levels at 02.00 hours. In the subjects on 0.71 or 1.5 gm of protein per kilo body weight peak plasma values were seen at or after 10.00

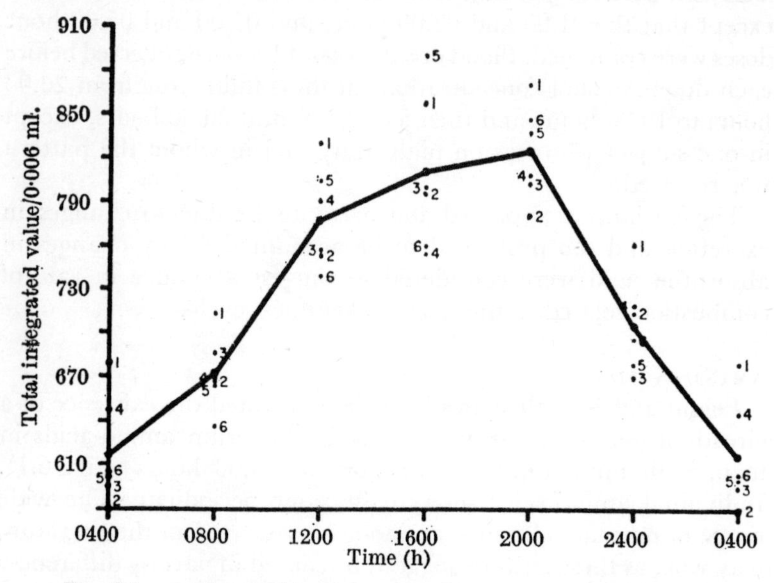

Fig. 6.1. Amino-acid levels in whole blood. The total integrated value of amino-acids/0·006 ml of whole blood has been averaged for each man and day and each point represents the mean of 30 determinations. The mean value for each man at a given hour is indicated by the subjects' numbers. (Feigin, R. D., Klainer, A. S. and Beisel, W. R., Fig. 1, 243.)

hours. In the subjects on less than 0·04 gm of protein, peak levels occurred rather earlier in the morning. A second peak in tryptophane concentrations was seen in the evening samples. The other thirteen amino-acids all showed a variation in concentrations also. The highest percentage changes were seen in tyrosine, tryptophane, phenylalanine, methionine, cysteine and isoleucine. The lowest percentage changes were observed with alanine, glycine and glutamic acid, all of which had relatively high concentrations even in the subjects on a low protein diet.

Studies (938) on one subject who was given protein meals at 3-hourly intervals showed less marked alterations in concentrations suggesting that frequent intake of protein could reduce the amplitude of the tyrosine rhythm but, since the rhythm persisted in subjects whether on a low or a high protein intake, it would appear that the periodicity in amino-acid concentrations is not due to protein intake. As Wurtman and his colleagues suggest (939), however, the rhythm could be connected with hormonal changes or dietary factors affecting the total size of the free amino-acid pool. The cortisol rhythm might be particularly important in this respect as glucocorticoids are known to lower plasma tyrosine levels (765) and the synthetic steroid dexamethasone has been shown to be capable of producing a phase shift in the plasma tryosine rhythm in man (940). The plasma insulin rhythm (266) could also play a part as insulin has been shown to alter the transfer rate of amino-acids across cell membranes (767).

Normal subjects given a standard dose of 3 g of tryptophane at 03.00, 06.00, 09.00, 12.00. 15.00, 18.00, 21.00 or 24.00 hours have shown a circadian variation in excretion of metabolites of the kynurenine pathway. Almost three times as much kynurenine, kynurenic acid and xanthurenic acid was excreted following tryptophane given at 09.00 than when it was given at 21.00 hours. Indican, a tryptophane metabolite not in the kynurenine pathway, varied inversely (740, 741). A circadian rhythm in the liver enzyme tryptophane pyrrolase, an enzyme in the kynurenine pathway which may control the rate of tryptophane metabolism and which has been shown in animals to have a circadian rhythm (741), could account for the circadian variation in tryptophane levels.

SERUM ENZYMES

Conflicting reports on the occurrence of circadian rhythms in serum enzyme activity have been published (262, 420, 577, 679, 842, 872).

In a study of 8 female and 16 male subjects a circadian rhythm in serum glutamic-oxalacetic transaminase with peak values about 18.45 hours and low values at midnight was reported (262). Other workers, however, although noting daily fluctuations, were unable to find a circadian rhythm in either the levels of activity of serum glutamic pyruvic transaminase or lactate dehydrogenase (577) or glutamic-oxalacetic transaminase (577, 872) or creatine kinase (679). A circadian rhythm in plasma lactic dehydrogenase levels has, however, been reported (842) and has been confirmed by Jacey and Schaefer (420) who obtained venous blood samples from 13 subjects at 4-hourly intervals. All subjects showed a circadian rhythm. The highest activity was seen in the afternoon, at 16.00 hours in six subjects and at 20.00 hours in the other seven. A circadian variation in erythrocyte glucose-6-phosphate dehydrogenase and 6-phosphogluconate dehydrogenase has also been reported (369).

ELECTROLYTES

IRON

Heilmeyer and Plötner, 1937 (371), noted an evening fall in serum iron in a single subject. This decrease was also noted by Valquist, 1941 (883), in a single 24-hour study. Further work has shown a circadian rhythm in plasma iron over the 24 hours with an early morning peak and low evening values (351, 371, 379, 407, 408, 698, 792, 798, 901).

In Høyer's very thorough study (407, 408) 12 normal subjects, six men and six women, were observed over a series of seven 24-hour periods. Samples were obtained at 08.00, 12.00, 16.00, 20.00, 23.00 hours and 08.00 hours the following morning in each series. Eighty-two 24-hour curves were obtained and of these 66 showed a marked fall (average value 42·7 μg/100 ml) in serum iron from morning till evening. Fifteen showed practically constant values throughout the 24 hours and only one curve showed a rise during the course of the day. Samples from a further 20 subjects over a 24-hour period from whom specimens were also obtained at night, confirmed this pattern and showed in addition that iron levels usually (15 out of 20 subjects) reached their lowest levels at night. Similar results were found in both males and females. In night workers peak values are seen in late afternoon or evening samples, i.e. in phase with their sleep/wakefulness cycle (351, 408, 598, 901). In subjects working at irregular hours, the rhythm is lost (351) and it can be disturbed relatively easily by changes in habit (341). It persists, however, in patients with adrenal insufficiency (341, 351, 598) and plasma iron concentrations are known to be unaffected by administration of cortisone (598), so that, although the plasma iron rhythm usually parallels the plasma corticosteroid rhythm, there does not appear to be a common controlling mechanism and the control of the plasma iron rhythm is as yet unknown.

Iron transport, as calculated by the disappearance rate of isotopic iron following sequential injections beginning at 01.00 and 13.00 hours, has been reported as showing a rate of 0·078 mg/hr/100 ml plasma at 01·00 as compared with 0·127 mg/hr/100 ml plasma at 13.00 hours (662). The same observers also noted a slowing of the initial rate of clearance occurring after 3 to 4 hours in the studies begun at 01.00 hours and at 8 or 10 hours in the series started at 13.00 hours. The authors suggest that the marrow uptake of iron may alter with time of day or that release of iron from reticuloendothelial cells into the plasma varies. In another study (546), however, no significant change in plasma iron transport between morning and evening was found. No variation during the course of the day has been observed in total serum binding capacity (93, 838).

COPPER

Nielson (675) noted a rise in forenoon values of serum copper. Munch-Peterson (655) performed serum determinations throughout the 24 hours and found a circadian variation with low values at night, a rise in the forenoon and then a gradual fall throughout the day. He also estimated serial samples taken from subjects on night duty and found some evidence of adaptation.

Calcium, magnesium, phosphate, sodium, potassium and chloride levels in blood are considered with their renal excretion in Chapter 4.

DIGESTIVE TRACT

SALIVA

A nocturnal fall in the pH of saliva normally occurs but in subjects who remain awake no decrease is seen (312).

A daily variation in concentration of a tyramine-like substance in human saliva, with low morning values and an afternoon peak followed by a fall in the evening, has been noted (350). The variation did not appear related to the body temperature rhythm.

DENTAL TISSUES

Periodontal connective tissue has been reported (388) as showing a 24-hour periodicity in tensile strength when judged by tooth-mobility. This was found to be lowest at night in the two subjects studied. Injections of ACTH or prevention of sleep abolished the rhythm.

There is some evidence that crevicular fluid flow may exhibit a circadian periodicity. In a group of fifteen healthy subjects, crevicular fluid flow was measured at 4-hour intervals over a 24-hour period beginning at 06.00 hours. A circadian pattern was seen with maximum flow at 22.00 hours (70).

GASTRIC FUNCTION

A rise in acidity of gastric juice during sleep together with a fall in volume was noted by Johnston and Washeim in 1924 (429). These results have been confirmed by many other workers (377, 788, 884, 928). Much individual variation occurs, however (256, 507); during the day some subjects show a rising acidity, others high morning levels with an afternoon fall. At night there is again much intersubject variation but in general secretion appears to be higher before 02.30 hours.

BILIARY FUNCTION

Twenty-four hour rhythms in serum bilirubin (38, 125) and in urobilinogen (269) have been described, as also a 24-hour rhythm in bile formation (255, 442, 714).

EXOCRINE PANCREATIC FUNCTION

Pfaff in 1897 (713) noted that there is a 24-hour variation in the flow of pancreatic juice. Excretion is usually at a maximum in the early afternoon and falls to a minimum at night (558), and it has been suggested (231) that its rhythm is determined by the entry of food and gastric contents into the duodenum and small intestine. Recently, however, it has been claimed that in subjects restricted to a water intake only the pancreatic exocrine function has a circadian rhythm (527). In the two patients studied, however, the only satisfactory appearance of circadian rhythmicity was in the flow rate, and this only in one subject and when he was receiving regular meals.

COLONIC ACTIVITY

A reduction or cessation of colonic activity between 22.00 and 04.00 hours and a resumption between 04.00 and 06.00 hours has been noted (19).

CELLULAR RHYTHMS

MITOSIS

A mitotic rhythm in human epidermis was reported by Cooper and Schiff (167) and this finding has been confirmed and further extended by Cooper (166) in a study of preputial skin samples. These were taken at circumcision in 57 infants aged 6-11 days. At least one specimen was obtained at every hour throughout the 24 except at 18.00 hours. Mitotic counts were made on 5000 or more cells in each sample. The results showed that mitotic activity was greater at night than during the day. A maximum was reached between 21.00 and 22.00 hours and a minimum between 05.00 hours and 10.00 hours. The highest count in a single specimen was 35 (out of a count of 5124 cells) in a piece of skin removed at 21.45 hours and lowest, 7 (out of 5000), in a specimen obtained at 10.25 hours. The lowest average mitotic count for a given hour was 9 at 05.00 hours. Broders and Dublin (107) and Dublin *et al.* (218) also noted higher mitotic rates in human epidermis from neonates in samples collected in the late evening as compared with those taken in the morning or afternoon. Scheving (795, 796) obtained samples of skin by punch biopsy from the shoulder region in young male volunteers. 193 biopsy samples were collected and mitoses in 2500-3000 cells from each counted. A circadian rhythm was observed with a peak in mitotic figures between 00.00 and 04.00 hours. 'Pre-chromosomal cells' were responsible for the early part of the peak, 'chromosomal' for the middle and 'reconstruction' cells for the last portion. A smaller early afternoon peak was also noted but did not appear to be statistically significant.

A similar circadian mitotic rhythm in adult human epidermis has been noted in another study (248). Peak numbers of mitotic figures were seen between 01.00 and 03.00 hours. There was a fall at 05.00 hours, a rise again at 07.00 hours and then a steady decline to 13.00 hours followed by a gradual increase in the afternoon and evening. Exercise was found to reduce mitosis almost to zero and the author suggests that the circadian mitotic rhythm is caused by a reciprocal variation in corticosteroid levels, and that during stress as from severe exercise an adrenalin effect is superimposed. Observations upon mitosis in mouse epidermis (129) indicate that such hormonal effects are not direct, but that adrenaline enhances the effect of the epidermal chalone, and cortisol prolongs the action of adrenaline. Whatever the mode of action, these mitotic rhythms may well result from the circadian rhythm in the adrenal cortex.

A clear circadian variation in mitotic activity of human bone marrow cells was found in five out of six subjects examined (578). Peak numbers of mitotic figures occurred at 24.00 hours and lowest numbers at 06.00 hours. There is thus general agreement about the timing of human mitotic rhythms.

Circadian variations in mitotic activity in human tumour cells have also been described (854, 899, see also chapter 11).

Mitotic rhythms must involve nucleic acid turnover; and DNA content and synthesis, cell proliferation kinetics, lyo- and dermo-enzymes, and growth processes have all been examined in animals and found to show circadian variations (120, 898).

THE EYE

INTRAOCULAR PRESSURE

The occurrence of daily variations in intraocular pressure appears to have been first noted by Hugeinin in 1899 (409) using tactile tension. Maslenikow in 1904 (575) using applanation tonometry took measurements at 09.00 and 17.30 hours and confirmed the existence of the phenomenon. Its circadian nature was indicated by Thiel (865) who made a series of records of the intraocular tension in five patients at 08.00, 10.00, 13.00, 17.00, 19.00, 23.00, 01.00. 03.00, 05.00 and 07.00 hours over four to five days. There is now a considerable literature on the subject. Most reports record maximum pressures between 06.00 and 08.00 hours and lowest values late in the evening. High nocturnal pressures have been noted by de Venecia and Davis (192) who found that in a group of 115 normal subjects the circadian variation was from 0 to 17 mm.Hg with a mean value of 5·9 mm.Hg. Drance (215), who made measurements in 404 normal eyes between 06.00 and 22.00 hours only, noted a variation of 0 to 10 mm.Hg with a mean of 3·7 mm.Hg, but had nocturnal readings been excluded from de Venecia and Davis's figures, they would have been very similar.

The mechanisms underlying circadian variations in intraocular pressure are not fully understood. Daily fluctuations in aqueous inflow have been noted (237) and these or changes in outflow resistance could be responsible (221, 529). Variations in episcleral venous pressure have been reported by some workers (37, 866, 867) but denied by others (528), an alteration in the pore diameter of the aqueous outflow paths has been suggested (176). Newell and Krill (671) have, however, found that daily changes in outflow facility are irregular and not in phase with the variations of intraocular tension.

A connection between adrenal cortical function and intraocular pressure may exist. This is suggested by the observation that circadian variations in intraocular tension are reduced in patients with adrenal cortical insufficiency (530) and the finding that maximum and minimum intraocular pressures appear to be related to plasma corticosteroid levels (95).

PUPILLARY SIZE

Daily variations in pupillary size have been noted with a marked morning increase followed by a fall during the afternoon (212). A circadian rhythm in the kinetics of the light reflex has also been reported in a group of seven subjects studied over 16 days. Maximum values were observed at 24.00 hours and minimum values at 15.00 hours (870).

Chapter 7

Wakefulness, Alertness, Skills, Time-Sense

Cerebral Function—Wakefulness—Performance Tests—Industrial Situations—Connection of Performance with other Functions—Time Estimation.

The title of this chapter is studiously vague. It might indeed be described as a rag-bag for a number of functions whose common property is that they are exerted by the higher regions of the central nervous system. Skills are many and varied, and cannot yet be analysed in terms of a finite number of completely independent abilities, contributing in a known manner to performance on the numerous tests devised to assess them.

Let us consider a familiar example. Most of us would accept that when we are sleepy, we are able neither to drive a car safely and accurately, nor to perform difficult intellectual tasks. It would also appear that different abilities are involved in these two tasks, since intellectuals are not notably good drivers, nor is a high level of driving skill a guarantee of high intellect. If it is found that these abilities vary circadianly, it remains an open question whether this results solely from variation in wakefulness, or whether one or more distinct abilities are fluctuating with a similar time course.

Both educational and industrial psychologists have been interested in devising tests which might predict a person's performance, whether this be competence at any particular industrial task, or ability to benefit from some form of higher education. Such tests differ fundamentally, however, from most physicochemical measurements in that it is uncertain precisely what is being measured, although certain broadly different categories of skill can be recognized.

Although difficult to measure and interpret, these skills are of the highest practical importance. When men are employed on night or rotating shifts, it is important that they should be able to operate machinery with the necessary skill, but it matters not whether their core temperature, or pulse rate, is appropriately adapted. When a statesman or business man flies halfway round the world to take part in delicate negotiations, he does not care whether his renal excretion of potassium is appropriately adapted, but he is closely concerned that his mental faculties shall be at a high level. One might indeed say cynically that useful variables cannot be measured, whilst those which can readily be measured are of little practical value.

CEREBRAL FUNCTION

A simple objective measurement of cerebral function can be obtained from the electroencephalogram (EEG), which is claimed (263) to show circadian variation in subjects following a nychthemeral existence. It has also been recorded every 2 hours in five subjects who remained awake and recumbent, and whose wakefulness was verified by simple tests of reaction time (71). In four of the five subjects, the frequency of the rhythm was lowest at 04.00 and 06.00 hours, which roughly coincided with the time when their reaction time was longest. The interesting rhythms in the sleeping EEG however have a much shorter period, of a hour or two.

Another simple measurable function of the central nervous system is the ability to detect a stimulus, measured as absolute threshold of sensation. Little useful investigation of rhythmic changes in absolute sensory thresholds is on record, partly because these are so prominently affected by the past history of experience of that modality of sensation. Absolute visual threshold, for example, falls for at least 12 hours when the subject is in complete darkness, which precludes frequent measurement.

A diurnal variation in auditory threshold has been suggested by some observers (669, 670) but denied by others (355, 908). Harris and Myers (355) examined thresholds at 256, 1024, and 8192 cps from 08.00 to 16.00 hours in a group of three subjects but found no significant variation. Similarly Ward (908) in a careful investigation in which he measured auditory thresholds at 500, 1000, 2000, 3000, 4000 and 6000 cycles per second in six subjects at 07.00, 12.00, 17.00 and 24.00 hours for several days, could find no evidence of a circadian rhythm.

Variations in olfactory acuity in the course of the day appear to be related solely to food intake (291). Threshold falls between meals but rises whenever food is taken. It is, however, reported that sensitivity to pain is highest in the evening, falling again to a low level throughout the night (441).

WAKEFULNESS

Subjectively the most dramatic circadian rhythm is the alternation between the consciousness of our waking hours, and the unconsciousness of sleep, though it has been suggested (876) that this is only due to social and occupational pressures, without which the sleep/wakefulness pattern would be polycyclic. 'Sleep and Wakefulness' is the title of a book by Kleitman (470) whose second edition contains five chapters upon periodicity. It might easily be supposed that there is here no endogenous rhythm, that we are sleepy after 16 hours or so awake, and wakeful again after 8 hours of refreshing sleep. We all know that many other influences can make us wakeful, such as anxiety, a high sensory input, or an active or interesting occupation, whilst we become sleepy through boredom, repletion, warmth, comfort, and limited sensory input. Anyone who has attempted critical self-assessment will, however, be aware that there is also a rhythmic circadian waxing and waning of sleepiness. Social gatherings which continue through the night tend to be less lively in the small hours of the morning, and more so as the morning advances past dawn; most people will admit to a time of night when they felt very sleepy, followed by increased wakefulness, even though they had not slept, and this applies equally whether the vigil was for pleasure, for work, or for experimental purposes.

A striking example is shown in Fig. 7.1 which represents unpublished data from an experiment in which soldiers were kept awake for three days and practised shooting every 3 hours. Their self-assessed fatigue increased sharply during the night hours and perceptibly diminished during the day, although they were screened from any indication of the passage of day and night (506).

The same phenomenon was dramatically illustrated for one of us in the course of a 24-hour mountain expedition. While there was difficulty, discomfort and potential danger, the party was not aware of sleepiness. By 05.00 hours we were walking back down a glen in daylight, and repeatedly fell asleep as we walked, to be awakened as we stumbled on obstructions. Though it was over 24 hours since our last meal, the urge for sleep was stronger than any appetite for food, and we expressed our firm intention of going straight to bed as soon as we were back in the hut, without delaying to have breakfast. When in fact we were back, about 09.00 hours, we felt quite different, and none of us was particularly anxious for bed until we had cooked and eaten an ample breakfast;

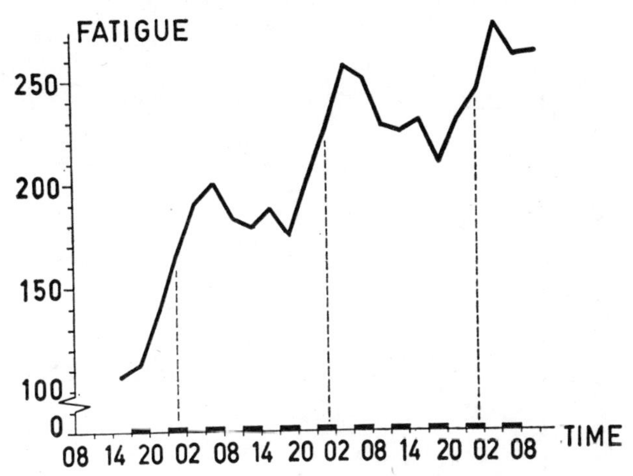

FIG. 7.1. Fatigue, subjectively assessed, in 63 subjects awake and active for 3 days. (Fröberg, J. and Levi, L., unpublished data.)

even then we slept for only 6 hours or so, although we were physically tired, very short of sleep, and in warm, comfortable bunks in a quiet, dark hut. Next night, however, our sleep was long and deep.

Similar evidence, that beyond the numerous other influences there is a circadian variation in wakefulness and sleepiness, comes from a variety of circumstances in which people have tried to live on abnormal time schedules. Kleitman (475) records how two subjects attempted to live on a 28-hour day, alone in a cave; one of them had no difficulty, whilst the other found that his presumably endogenous rhythm failed to adapt, so that he had great difficulty in sleeping when the artificial day was out of phase with real time although all conditions were favourable, including the presence of a companion who was sleeping soundly. This observation illustrates that some individuals can adapt their sleeping habits much more easily than others, as has also been observed during life on a day of abnormal length in an Arctic summer (101, 517). Some difficulty has also been recorded in subjects reversing the phase of their habits (603) or living on a 12-hour day (613). The difficulty usually experienced by night workers in sleeping during the day cannot so easily be ascribed to inherent rhythmicity, since many conditions are unfavourable, such as noise, daylight, and perhaps the psychological effect of attempting to sleep when the world around them is active.

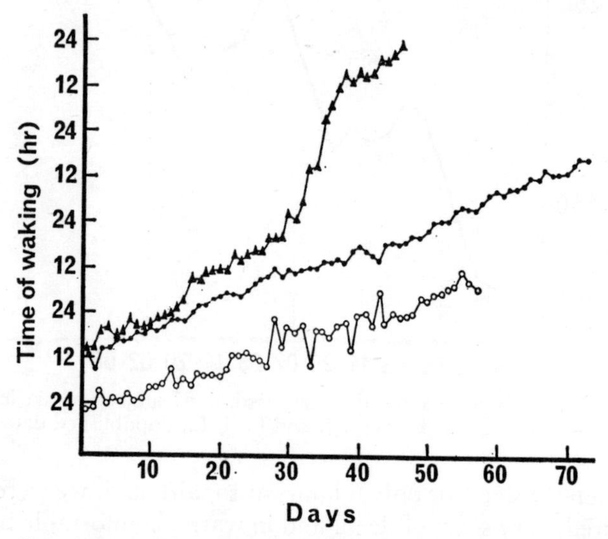

FIG. 7.2. Time of waking of three subjects alone in caves. ○ Siffre. ● Workman. ▲ Lafferty, latter half of sojourn.

Human subjects may be freed from social and climatic constraints upon their sleeping habits by living in isolation underground. Many observations upon the times of waking and sleeping of such subjects have now been recorded, both in caves (153, 261, 346, 607, 609, 611, 818, 819) and in an underground bunker (30, 31, 32, 922). All agree that in such circumstances men assume habits of waking and sleeping which indicate a day somewhat over 24 hours (Fig. 7.2). Estimates for the cave-dwellers studied by different authors are:

	Hours
Siffre	24·5
Workman . . .	24·7
Laures	24·6
Senni	24·8
Lafferty . . .	25·0
Maretaet . . .	24·6

over periods of from 8 to 25 weeks. Aschoff's subjects in the bunker showed somewhat wider variation, perhaps because most of them were only studied over periods of 3 weeks or so. Most showed

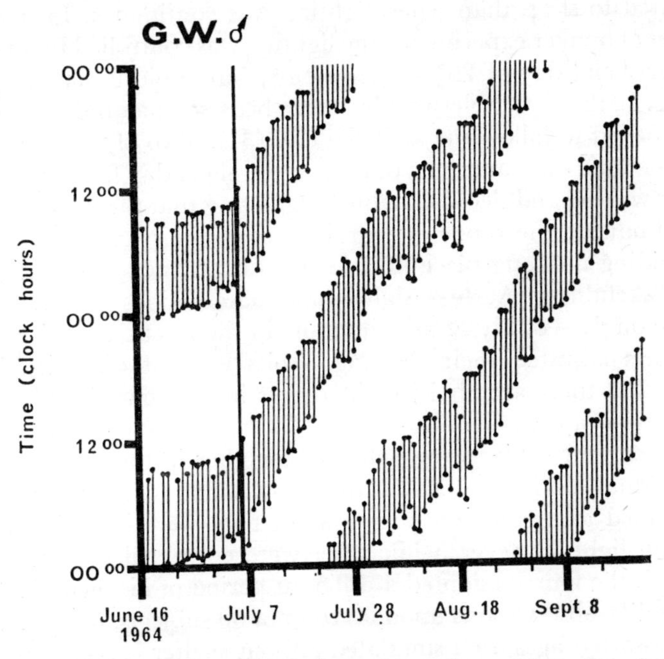

Fig. 7.3. Continuous lines indicate times when subject, alone in a cave with a watch, slept; before the vertical line he attempted to sleep at his customary hours, subsequently he went to bed and got up when he felt inclined. The ordinate spans 48 hours, so each sleep period is represented twice. (Figure drawn under control of an electronic computer programme, by kindness of F. Halberg.)

periods of slightly over 24 hours, as did the cave dwellers; but two unusual subjects lived on cycles lasting 34·4 and 42·5 hours, a pattern shown also by cave dwellers over limited periods of time.

A series of 75 such bunker experiments has now been recorded, with varying levels of illumination, and including some where the room was continuously light and others where the subject could switch off the light when he wished to sleep (922). These have failed to confirm the consistent effects of varying illumination upon other species, summarized in Aschoff's Rule (23, 24), which states that in a nocturnal animal such as a mouse the period is lengthened in bright and shortened in dim light, whilst in a diurnal animal such as a finch it is lengthened in dim and shortened in bright light. It seems that man is less directly affected by illumination level than are other species. The period was, however, significantly longer when subjects could switch off the light when they wanted to sleep than when lighting was continuous. In another series of bunker experiments the lighting was controlled to produce an artificial day of $26\frac{2}{3}$ or $22\frac{2}{3}$ hours, but subjects had reading lights both at a table and by their beds so that they were not compelled to follow the artificial day (31). Two of seven subjects on the long day, and four of five on the short day failed to adjust their waking and sleeping habits, suggesting that these day lengths were outside the range of entrainment of their habits, and thus providing a strong piece of evidence for an endogenous rhythm in wakefulness. At first sight these observations conflict with those on days of 21, 22 and 28 hours in the Arctic to which most subjects adjusted their sleeping habits without difficulty (101, 517); but there are obvious differences in the conditions in these two series of experiments.

When four subjects were studied together in the bunker, their individual periods, derived from isolation experiments, were entrained to a common period when they were in bright light; in dim light, however, while three were entrained to a common period, the fourth adopted a different period of his own (725).

Similar observations upon a group of 30 subjects, of both sexes and assorted ages, in a simulated fall-out shelter (170), showed a spontaneous periodicity of habit of about 24 hours. This experiment only lasted however for 13 days, and individuals were constrained to follow the habits of the group; and no periodicity analysis was performed to determine whether the period departed significantly from 24 hours.

Some individual peculiarities are worthy of record. Workman (607) had a watch with him, and was intending to follow normal time, but after a week or two of isolation he found this increasingly difficult, being unable to get to sleep at night, and sleeping late in the morning; after 3 weeks he abandoned his intention, and went to bed when he felt inclined, thus adopting the 24·7 hour cycle mentioned above (Fig. 7.3). Knowledge of the time was inadequate as a Zeitgeber, and it would seem that an endogenous rhythm of around 25 hours forced him to modify his intended habits.

None of the other subjects had a timepiece, and their habits of waking and sleeping were recorded by a signal transmitted to the surface. Maretaet, Lafferty, and some of the subjects of Aschoff at some time lived on days of around 48 hours or slightly over, suggesting that if they did not go to sleep at the expected time, they did not feel sufficiently sleepy until another cycle had elapsed.

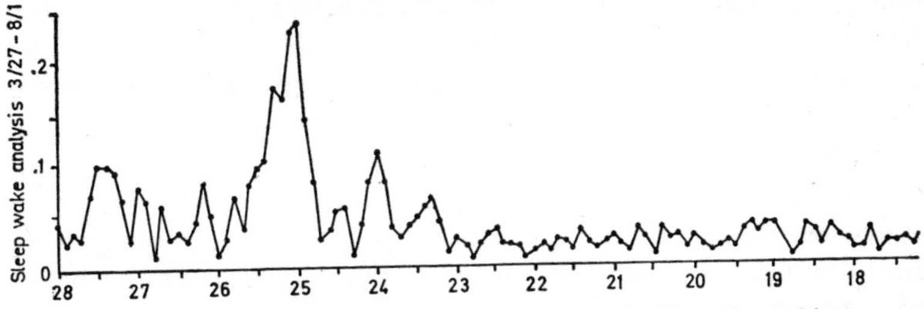

Fig. 7.4. Spectral analysis of sleep-waking habits of a subject alone in a cave for four months. Each point represents the amplitude of the best-fitting sine curve of a period of length indicated on the abscissa. The peak, around 25·0 hours, stands out clearly from the 'noise'. (Drawn under control of an electronic computer programme by courtesy of Professor F. Halberg.)

Lafferty's habits, especially for the first half of his stay, were grossly irregular, and casual inspection reveals no discernible rhythm. If, however, a series of sine curves of slightly different frequencies, with periods at intervals of 0·1 hour, are fitted to the data (kindly performed for us by computer by Professor F. Halberg), a period of 25·0 or 25·1 hours clearly provides a much better fit than does any other period (611, Fig. 7.4). This presence of a statistically significant rhythm in apparently irregular data has important implications for the interpretation of sleep data. The time when a man goes to bed is affected, even in isolation, by many influences beyond his circadian wakefulness cycle, such as fatigue, warmth, and interest in or boredom with what he is doing; in fact he makes a free choice, only determined in part by his endogenous rhythm which imposes certain moments of maximum likelihood, just as a man may die or be born at any hour of day or night, though there are certain hours of maximum likelihood. One need not infer that a subject such as one of Aschoff's (30), living on a 34·4 hour cycle, has an endogenous wakefulness rhythm of this period, particularly if the observations are not prolonged. It is, therefore, desirable to look for some other criterion of the activity of the endogenous clock, beyond a mere record of the times of sleeping and waking.

PERFORMANCE TESTS

Most of those who have devised and operated tests of alertness or various motor, sensory and mental skills have not been primarily concerned with rhythmic changes. Concern over the reliability of test results led however to the recognition that they can be affected alike by practice and fatigue, as well as by events in the course of the day; a post-prandial decline is common (274, 275, 496, 656, 942). The expected decline through fatigue is unpredictable and may even be reversed as the subject 'gets into the swing' of testing procedures; some such explanation is needed for the observation that candidates performed better at the end than at the beginning of a highly motivated 3-hour testing session (676). We have ourselves made similar observations of the dramatic effects of a seemingly trivial change of routine.

It is also commonly found (121, 942) that in circumstances where performance declines, it also becomes more erratic, as if the subject by a special effort could produce a good performance, but was unable to sustain it at this level.

If rhythmic variations in test results are to be studied the test must be capable of extensive replication. This presents no difficulty, outside that of improvement through practice and decline through fatigue, in such simple tests as reaction time; but for the more complex 'problem' type of test it is necessary to construct a series of problems of equal difficulty. Attempts to construct and assemble such tests are now being made (642, Fort, A. and Mills, J. N., unpublished data). Since it is not known how many independent functions or abilities can be recognized, no entirely logical classification of different forms of test is possible. All tests involve a sensory, and usually perceptual, input and some form of motor output, with an intervening cerebral process, so any division into perceptual, motor and mental tests depends upon an assessment of which aspect of the task is the most difficult. A test may involve a difficult sensory discrimination, followed by a simple answer; it may involve a simple instruction followed by an accurately controlled movement, such as maze-tracing, or inserting dots in small circles; or it may involve lengthy thought between the reading of the problem and the writing of the answer. The distinction is not, however, always clear-cut; the components may all be easy and rapid as with simple reaction time, or more than one may present difficulty, as with motor track simulators. Rapid arithmetic may be intended as a mental test, but if the mental process is sufficiently quick the time taken will depend in part upon speed of reading and writing. Difficulties of assessment may arise when the subject is asked to work rapidly, answering as many questions as possible, and one has to score both for speed and for number of mistakes.

A special variant of a primarily perceptual test is the measurement of so-called 'vigilance' (556). Here the subject is required in the course of a long session, at least half an hour and preferably longer, to pick out and respond to an occasional signal, visual or auditory, from an irregular background of sensory input. Neither the individual recognition of the signal nor the response to it is a difficult task, but it requires constant attentiveness. This test has obvious and direct implications in industrial and military situations, for those who have to monitor instruments or radar screens.

A quite different and purely subjective method of assessment is to obtain answers to a questionary (352, 864). For instance, the subject is offered a list of adjectives and is asked to rate himself, on a four-point scale, in terms of how far the particular adjective describes him at that moment.

Many studies have been confined to the working day, and do not therefore offer much evidence for inherent rhythmicity owing to the great variety of rhythmic influences. Kleitman (470) reviews much of this work, including observations of his own, which indicate a peak performance in a number of tests, mainly mental or motor, around midday (Fig. 7.5). A very similar peak has been found in measurements of simple reaction time in adults (8, 464) and in children (352), in the state known as 'activation' derived from a self-assessment questionary (864), and in mental tests such as quick arithmetic and memorization (274, 275). Minor variations in timing such as a somewhat later peak (464, 496) may represent merely the different populations studied; and it has been claimed (396) that the timing is adapted to the subject's customary hours of work, the peak being earlier in those who started work at 07.30

hours than in those starting at 10.30. Motor skill, as in maze-tracing, has been found to rise in the afternoon when mental skills were falling (275, 396), suggesting that there is more than one fluctuating influence. A somewhat different timing was found in a group of 25-30 naval ratings, whose performance on most

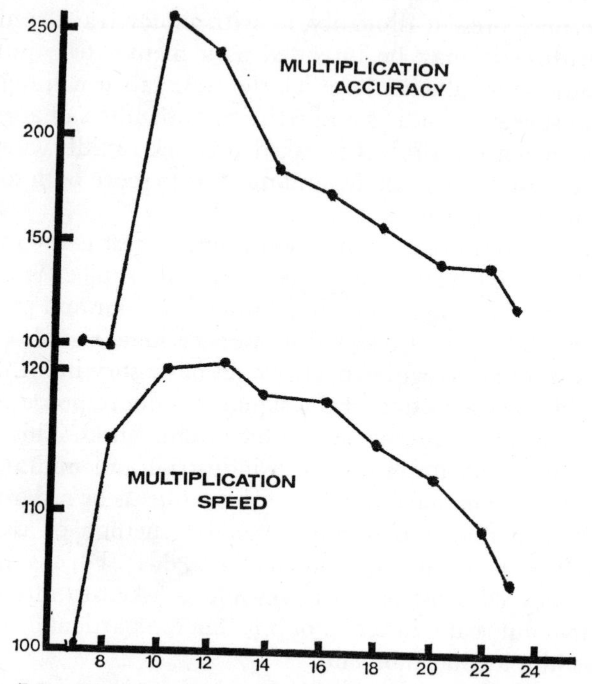

FIG. 7.5. Circadian variation in speed and accuracy of multiplication as percentage of 07.00 hour value. Mean of 20 days on one subject. (From Kleitman, N., Fig. 16.1, 470.)

tests improved steadily up to 21.00 hours (74, 76). This may again reflect a different population, but, in addition, most of the tests were relatively simple but tedious, and in the one with the highest mental content, memorization, the performance was best at 10.30 hours and declined steadily thereafter.

When observations have been made by night as well as by day, it is always possible that subjects perform poorly simply because they have just been awakened from sleep. This specific point was investigated by Webb and Agnew (910) who put subjects to sleep after lunch and waited until their electroencephalogram indicated

Stage 4 sleep before waking them and testing them. They found that recovery was complete in 5 minutes, and probably within one minute. The test which they used was, however, a very simple key-pressing task, and recovery of performance on a more complex task might well be slower, as Jansen and co-workers claim (423). These workers tested two subjects on 12 days, at 2-hourly intervals through day and night. They used a tracking test simulating the driving of a car, in which the subject drove as rapidly as possible without hitting the verges; speed varied in a roughly sinusoidal pattern, the subjects driving fastest shortly after midday and most slowly around or shortly after midnight, but their accuracy followed a very different course, being best around midnight, and worst about 08.00 hours; they also did a mental arithmetic test, which they performed fastest around midday. Fort (258) used a perceptual-motor task in which subjects were required to scan a block of letters and cancel any which were identical with the preceding letter, the score being the number scanned in a minute. In the course of 6 days of testing, the test was applied twice at every hour of the 24. Most subjects performed poorly in the small hours of the morning, and their performance improved sharply to a steady or irregular plateau during the day. Simple reaction time (8) has also been found to be slowest at 04.00 hours, and during 3 days without sleep a rhythm in speed of arithmetic persists, performance being better at 20.00 than at 08.00 hours. Performance on a simple motor test followed a circadian rhythm in subjects living a 48-hour day (579).

Experiments in the authors' laboratory (unpublished) have compared performance upon a more 'intellectual' task, the separation of false from true syllogisms, with simpler perceptual and motor tests such as have been used by other workers. Each of 12 subjects began at a different time and was tested at 2-hourly intervals for 24 hours, without sleep, thus performing the battery of tests 13 times. For each subject, the last tests were thus at the same time of day as the first; and since performance was almost always better, practice was more important than fatigue. Since, however, the mean performance at each time of day was derived from the first and thirteenth attempt by one subject, the second by another, the third by another and so forth, any practice effect in the group was cancelled without the need to fit to it an algebraic function.

Performance on the 'intellectual' task improved until around midday and then declined. The improvement during the morning was paralleled by the rise of body temperature, but temperature remained high until the late evening, while test performance was declining. A test of motor coordination followed throughout the 24 hours a course similar to that of body temperature.

INDUSTRIAL SITUATIONS

Other studies have been made of performance in an actual job situation, best achieved with a process which continues, with relays of shift-workers, throughout the 24 hours. In a study on teleprinter operators (121) the delay in response to a call was longest between 03.00 and 04.00 and least in the early afternoon. There is also a very thorough investigation of errors made by meter readers at a gas works over a period of 30 years (72). The men worked 8-hour shifts, rotating, so that the same subject contributed values evenly spread throughout the 24 hours, and the maximum incidence of errors, around 03.00 hours, was related only to clock time and not to whether the man was near the beginning or end of his shift. The pattern was remarkably similar in each of the three men involved, as also at an earlier period when they worked a 7-day week and a later one when they only worked a six-day week.

These studies clearly demonstrated rhythmic variation in skills independent of the subject's pattern of life, although the subjects were living in a nychthemeral society. Complete isolation from such influences was achieved in experiments reviewed by Alluisi and Chiles (11, 141), representing some ten years' work which has mostly been published in official U.S. Government reports. The most striking results were obtained on subjects spending 30 days on either a routine of 4 hours test, 4 hours rest, or one with only 2 hours rest between the 4-hour test periods, and subjected to a battery of six tests, vigilance, sensory-perceptual, and procedural. They comment that 'a physiologically determined diurnal rhythm is present and underlies all performance'. Their primary object was not to search for circadian rhythmicity, but to attain maximal satisfactory performance, so the subjects were deliberately urged to try to overcome the circadian recurrence of poor performance periods; in this the subject was usually successful, unless he was overloaded with tasks or was short of sleep. Even though the subjects had departed from a nychthemeral existence for 30 days, a circadian temperature rhythm persisted, with maximum about

20.00 and minimum about 08.00 hours, though the rhythm was flattening in the last week.

This persistence of a circadian rhythm after the exclusion of environmental influences with a similar rhythm is in contrast to another investigation, in which three men spent 14 days confined to an experimental chamber, but living otherwise on nychthemeral habits, sleeping from 00.00 to 08.00 hours (265). Their performance on a vigilance and a response time test showed rhythmic variations, but despite the regularity of their 24-hour habits their performance rhythms showed periods of between 20 and 30 hours. We know of no other comparable observations of apparently free-running human rhythms beyond those made upon subjects alone underground.

CONNECTION OF PERFORMANCE WITH OTHER FUNCTIONS

We must now consider how far these variations in skills can be explained in terms of other functions which vary circadianly.

Halberg (325) maintains that the early morning secretion of cortisol, which is commonly observed before the subject wakes, is an alerting mechanism, but no-one appears to have investigated whether cortisol would improve performance at some time of day when it is poor.

It has often been pointed out (76, 470, 551) that the diurnal rhythm in performance closely parallels that in body temperature. Fig. 7.6 taken from Kleitman's data is particularly striking. Such coincidence is by itself of little significance when so many functions vary circadianly. It is commonly supposed (352, 864) that people can be divided into 'morning' and 'evening' types, perhaps corresponding (74, 75) to introverts and extroverts, whose temperature peak is achieved earlier and later in the morning respectively. It is also claimed that introverts achieve a high performance level earlier in the morning than do extroverts, the performance level being thus associated with body temperature (74, 154). In comparing different methods of scoring performance, Blake (76) observed that the sensitive index which was correlated with body temperature was speed rather than accuracy, as was also observed with a motor track simulator (423).

Further evidence that the association between temperature and performance is more than fortuitous comes from a recent study (155) in which temperature and efficiency at mental tasks were measured over 12 consecutive days on men on rotating or stable shifts. Three different tests of performance were used: frequency of detection of an auditory signal on a noisy background, latency of response to this signal, and speed of adding 2-digit numbers. The subjects on rotating shifts worked, in each three

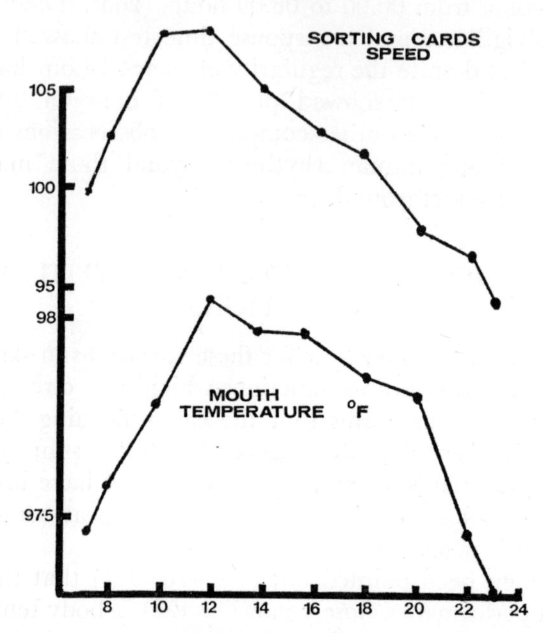

FIG. 7.6. Circadian variation in mouth temperature, and in speed of sorting cards, percentage of value at 07.00 hours. Mean of 20 days on one subject. (From Kleitman, N., Fig. 16.1, 470.)

days, each of the possible six 4-hour shifts in the 24 hours and were thus unable to follow either a nychthemeral or any other regular pattern of existence; but temperature and performance oscillated with rhythms similar in general form, with peak levels around 18.00 hours. The subjects on stable shifts worked from 00.00 to 04.00 hours and 12.30 to 16.30 hours, and at first their performance was poor on the night shift, when their temperature was low. Over the 12 days of the experiment, however, their night performance and temperature improved more or less in parallel, so that by the end the nocturnal fall in temperature had been delayed by about 4 hours and performance in night shift was similar to that by day. Thus on both systems body temperature was a predictor of performance efficiency. The same holds good (156) in men working on more conventional 8-hour shifts by day, by nights and in the early morning.

If one accepts that the association between temperature and performance is so close as to suggest a causal connection, there remain the two possibilities, that high internal temperature directly improves performance, or that it is merely an indicator of the state of arousal, which is said specifically to improve motor performance.

This could only be answered with confidence by deliberately raising body temperature and looking for an improvement in skill. Such a result is in fact claimed by Kleitman, who observed that lying down caused a fall, and standing a rise, in body temperature (472), and further that reaction time diminished and increased accordingly (477).

Another study on the direct effect of raising body temperature, in which 12 subjects were studied at 37·3, 37·9 and 38·5°C, showed a decline in both the speed and accuracy of mental arithmetic as temperature rose, but improvement on an auditory vigilance test (260, 926). If, however, attention is confined to the temperature range likely to be encountered in the course of a nychthemeron, 'normal' to 37·3°C, speed of addition and signal detection both increased with rise of temperature, errors in addition hardly increased, and the reaction time in vigilance test was notably prolonged. Only this last is in conflict with the association between temperature and performance, on essentially similar tests, mentioned above in men on abnormal time schedules (155). The two phenomena, the association of performance with the circadian rhythm of temperature and with temperature artificially changed, should be studied upon the same subjects.

The association between body temperature and skill has been assumed by some workers to be so close that they have used temperature measurements to assess how well adapted are men working in industry on different arrangements of shifts (77, 137, 911).

Some doubt upon this interpretation is cast by observations of Aschoff (29) who postulates a much less direct connection between temperature and wakefulness as a result of studies on subjects isolated in an underground bunker. If they are deprived of knowledge of the time all their observed rhythms free-run with the same period, slightly over 24 hours, but the phase relation between temperature and the habits of sleeping and waking is altered. It is suggested that two hypothetical oscillators, controlling temperature and wakefulness, have become coupled in a phase relationship different from that observed when both are under the influence of external Zeitgeber. A complete dissociation between body temperature and activity rhythms in subjects in isolation is described in Chapter 9.

A further difficulty in supposing that all performance rhythms are closely dependent upon body temperature stems from the different circadian patterns which have long been known for different kinds of tests (396). It is claimed that motor performance improves during the course of the working day while mental abilities steadily decline and coordinative processes first improve and then deteriorate but, as indicated above, many reports fail to distinguish between circadian rhythm in performance, and the complex effects loosely described as practice and fatigue but including all possible effects of previous testing sessions.

TIME ESTIMATION

A different cerebral function which has been much studied by psychologists (150), but is not obviously related to the skills so far considered, is the ability to estimate intervals of time. This ability may be assessed quantitatively in three ways: estimation, whereby time spans are produced and the subject is asked to estimate their duration; production, whereby the subject is asked to indicate the passage of a stated interval of time; and reproduction, whereby he is asked to reproduce a presented time span. Only this last obviates the need for language and familiarity with the conventional spans of seconds and minutes. It is probable that the methods used are different for time spans of different length: seconds, minutes, hours or days, and with the longer times it is impossible to dissociate any direct sense of time from such adventitious clues as tiredness, hunger and so forth. Consistent differences also arise through such seemingly irrelevant factors as whether a time span is presented visually or auditorily (518). Time spans estimated or produced may be 'empty', or may be 'occupied', by, for example, attempting to tap or count once a second. It has been supposed that when body temperature is high, or the thyroid overactive, the process whereby one attempts to estimate the lapse of time is speeded up; hence if one tries to produce a span of a minute, one produces a shorter span, and a real minute is perceived as a longer period (462). This reciprocal relationship, a tendency to produce or reproduce too short an interval when the duration of a presented span is overestimated, does not always apply. During his 6 months underground, M. Maretaet (261) produced and reproduced progressively longer spells while his estimation remained remarkably accurate. Estimation is more likely to be vitiated by the subject's attempts to guess what interval might be presented.

The technique most commonly used has been the ability to count seconds; subjects have been found to count more rapidly when body temperature was raised by diathermy or in fever (390, see also 717) and more slowly when they were cooled (33). The relationship between speed of counting and body temperature was in fact used to calculate the Q_{10} for the supposed underlying chemical process.

Attempts to confirm these observations (55, 56) cast doubt upon a simple chemical explanation of the phenomenon. Different subjects, intending to tap at one or three per second, and warmed by immersion of the legs in hot water, increased their tapping rates to a very different extent; some indeed tapped more slowly although the mean rate was higher in warmed subjects. Errors in estimating the passage of a given interval of time, or the frequency of a metronome, showed no dependence upon the subject's temperature.

If, however, the findings of these different workers are put together (33), a clear relationship between body temperature and time sense emerges, with about 10 per cent speeding up for a 1°F temperature rise.

Most workers who have studied circadian variation in time sense have found the expected relationship with body temperature, whether they have used production or estimation, and whether production has been of empty time, or has been assisted by attempting to tap or count once a second. Contrary reports are of lack of rhythm in a group of 30 subjects (76), and of a phase relationship the reverse of that expected (869): two subjects said to be 'night-active' produced their shortest intervals at 15.00 hours, when four subjects said to be 'day-active' produced their longest intervals. Temperature was not measured, but blood pressure was higher at the time of supposed activity, when time production was slow.

In three reports of extensive measurements upon single subjects the expected phase relation has, however, been found. In two of these (323, 844) peak temperature was 12 hours out of phase with slowest production of a time interval, while in the third, where temperature was not measured, counting was slowest during the habitual sleep period (347). A much larger series, comprising 36 subjects and nearly 600 days, has recently been analysed, and though temperature is not mentioned, the slowest production of time, by attempting to count seconds from one to 120, was recorded around one or two hours before the middle of the habitual sleep period (315, 316). In another recent series ten subjects were asked to produce a span of 15, 30 and 60 seconds of empty time, and to estimate the duration of a span which in fact was 10, 20 or 30 seconds (712). The expected relationship was shown in both tests, for when temperature was highest the subjects produced too short a period, and overestimated the duration of a presented interval.

Much evidence thus suggests that time sense depends upon a clock which accelerates and slows with rise and fall of body temperature; but before we are confident that this is the sole cause of circadian variation in time sense, it would be valuable to have measurements upon the same subjects with deliberate, and with spontaneous circadian, changes of body temperature, to provide assurance that the amplitude of the circadian change in time sense is appropriate to the amplitude of the temperature rhythm.

A rhythm in time estimation in free-running conditions has been demonstrated (346) with a dominant period of 24·6 hours, phase-synchronized with the pulse rate (329). As all other measured physiological rhythms had assumed the same period, the rhythm in time estimation could have resulted from any of a large number of other physiological processes.

When we consider longer periods, it is abundantly clear that the timing device controlling the various physiological circadian rhythms is quite independent of subjective assessment of the passage of a day. All those who have spent months underground without a timepiece have underestimated the passage of time. Siffre, for instance, when settling off to sleep at intervals close to 24 hours, often thought that he was taking a brief post-prandial nap after his midday meal. Maretaet, though forewarned of this possibility, for a considerable part of his underground stay behaved in exactly the same way, spending up to 10 hours asleep when he thought he was only taking a brief nap after a midday meal.

Since spells of time of 24 hours and upwards are inevitably 'occupied', their estimation would seem to depend upon a host of psychological factors concerned, *inter alia*, with the nature of the occupation. By contrast, the 'clock' which controls physiological rhythms, if one can apply evidence derived from much more lowly organisms, relies upon a specific biochemical mechanism involving RNA and protein synthesis and perhaps a sequential linear transcription of the genome (223, 359). A simple example of such a 'clock' has been demonstrated in a single neurone in the invertebrate Aplysia (851), whose excised eye also discharges with a circadian rhythm (421).

The nature of the clock, since it is at present completely in. accessible to investigation in man, is beyond the scope of this book- It has, however, been examined in lower animals and more recently in man (921) by attempting to describe observed behaviour by a mathematical model. Also outside our scope is another aspect of central nervous function which can only be studied in other species, the concentration, in different regions of the brain, of amines supposed to be involved in transmission processes. Their concentration may fluctuate regularly over the 24 hours (268), and thus may be involved in an early stage of the transmission process from 'clock' to 'hands'.

Chapter 8

Circadian Rhythm in Birth. The Development, Synchronization and Maintenance of Circadian Rhythms

Circadian Rhythm in Birth—The Development of Circadian Rhythms in Sleep/Wakefulness, Heart Rate, Electrical Skin Resistance, Eosinophil Levels and Activity Rhythm, Corticosteroid Concentrations, Temperature—Synchronization and Maintenance of Human Circadian Rhythms: Alternation of Light and Darkness, Social Factors and other possible Zeitgeber, Dissociation of Rhythms, Endogenous Circadian Clock, Rhythmicity at Various Levels, Possible Underlying Mechanisms, Oscillatory and Electronic Models and the Chronon Concept.

This chapter is concerned with circadian rhythms in birth, their development in infancy and childhood, and their control and maintenance in adult life.

BIRTH

The presence or absence of a periodicity in human birth has been a matter of dispute for many years. As long ago as 1848 Danz and Fuchs (175) reported an analysis which seemed to show that births were more frequent during the night hours. There are, however, many difficulties in analysing such data, and even as recently as 1952 there was a controversy in the pages of the *British Medical Journal* over whether or not more births occur at night (821). The evidence would, however, appear now to be very much in favour of the existence of a quite marked circadian rhythm in human birth (140, 427, 447, 448, 459, 460, 479, 685, 724, 764, 943).

Charles (140) analysed the times of delivery for over 16,000 live and still births in an English city in 1951 and 1952. Excluding all induced onsets or deliveries, or any with conditions thought likely to affect noticeably either of these, 55·6 per cent of deliveries occurred between 21.00 and 09.00 hours with a peak of 03.00

hours. 55·5 per cent of domiciliary births took place at night as compared with 54·1 per cent of institutional deliveries, and in the peak 6-hour period, 30·3 per cent of domiciliary births occurred as against 27·9 per cent of institutional.

An analysis of 38,668 hospital deliveries in Oklahoma City over a 5-year period showed that in every month of the year the lowest incidence was between 15.00 and 16.00 hours. Comparison with the hours of daylight revealed that over the year as a whole 9·6 per cent more babies were born at night (724).

An analysis of 33,215 births from five hospitals in three different cities showed that 54·9 per cent occurred between 00·00 and 12.00 and 45·1 per cent between 12.00 and 24.00 hours. The peak hour was 05.00 with 48 per cent more births than at the lowest hour—19.00 hours (459). In a comparison of two groups of patients in a different city, a markedly different peak time for births was found in a Caucasian group of 10,469 deliveries, whereas in a Negro group of 13,266 in a different hospital the incidence at different times was similar to that above; but further investigation showed that induction of labour by artificial rupture of the membranes was a common practice in the Caucasian group and fully accounted for the apparent difference (460).

A most extensive survey has been performed by Kaiser and Halberg (448) who reviewed previously published observations, including those above, on the time of delivery in over 600,000 spontaneous births. A peak in birth frequency occurred between 03.00 and 04.00 hours with a trough in the late afternoon and evening between 13.00 and 20.00 hours (Fig. 8.1). Little information appears to be available as to whether the mothers were admitted to hospital before or after labour had begun, or whether or not they were ambulant. The influence of posture is thus not known.

The time of onset of labour has also been found to show a circadian periodicity with maximum values between 02.00 and 03.00 hours and a minimum about 13.00 and 14.00 hours (140, 276, 559, 560, 563, 816), and this may likewise result from postural changes.

Records of over 4000 births in a New York hospital (816) showed that in 79·75 per cent, labour began in the presence of intact membranes and in 20·25 per cent spontaneous rupture of the membranes occurred first. Plotting the time of occurrence of the two phenomena revealed a very similar circadian rhythmicity in both with peak values between 02.00 and 04.00 hours and lowest values in the early afternoon.

As with time of birth, distinctions are rarely drawn between onsets commencing outside hospital and those beginning after admission; Charles has, however, noted that in domiciliary

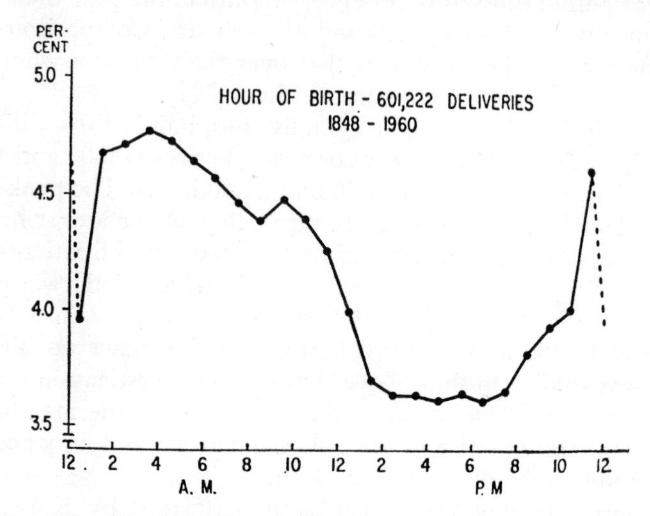

Fig. 8.1. Hourly incidence of 601,222 human births terminating spontaneously. (Kaiser, I. H. and Halberg, F., Fig. 2, 448.)

births a much more prominent peak in time of onset is seen and that normal labour seems to begin more frequently when the mother would normally be asleep (140). Labours beginning at night have been recorded as having a shorter duration than those starting during the daytime (559). This confirms the supposition that ambulance is a contributory factor.

LACTATION

A 24-hour periodicity in human lactation with a marked rise in milk output in the first morning feed has been reported (562).

DEVELOPMENT OF CIRCADIAN RHYTHMS

The course of development in infancy of a large number of circadian rhythms has been conveniently summarized (374).

SLEEP/WAKEFULNESS

Infants in the first few weeks of life show a 50 to 60 minute rest-activity cycle, and a total period of wakefulness of 8 to 9 hours out of the 24, with a range of 5 to 12 hours (286, 469, 470, 473). Both the timing and the duration of the sleep/wakefulness cycle in infants is different from the adult pattern (693, 694, 695, 696, 697).

Exogenous factors are considered by Kleitman to be predominant during the first three months of life and endogenous during the following trimester. The exogenous phase is governed by such external influences as the pattern of living in the family, temperature and noise levels and reduced willingness of parents to feed an infant during the night. During the second three months, the cerebral cortex has matured sufficiently for the infant to be more aware of his surroundings and able to communicate visually with his mother. This increasing awareness and the development of a steadier rhythm in the sleep/wakefulness cycle are accompanied by an increasing prominence of the alpha frequency of the electroencephalogram which ultimately results in a circadian rhythm with minimum rates between 04.00 and 05.00 hours. The waking hours between 08.00 and 20.00 hours increase and a circadian sleep/wakefulness and temperature pattern is established. In an infant allowed a completely permissive schedule and observed from the 11th to the 182nd day of life, there was an initial 25-hour sleep/wakefulness periodicity. By fifteen weeks this had become a 24-hour rhythm but with most of the sleep period between noon and midnight. At six months the sleep/wakefulness cycle was similar to that of other infants at that age (469, 471, and see Fig. 8.2). By the second year of life the day/night rhythm of sleep/wakefulness is fully established (372).

HEART RATE

Hellbrügge and his colleagues have observed the pattern of the heart rate in infants. During the last trimester before birth and for the first few weeks after delivery, the heart rate is independent of day or night, but from the 6th week onwards, daily changes in rate can be seen (373, 374, 375, 376).

ELECTRICAL SKIN RESISTANCE

Rutenfranz and his colleagues (780, 781, 782) have noted that the electrical skin resistance exhibits a circadian rhythm even in the first week of life and indeed, apart from some alteration in

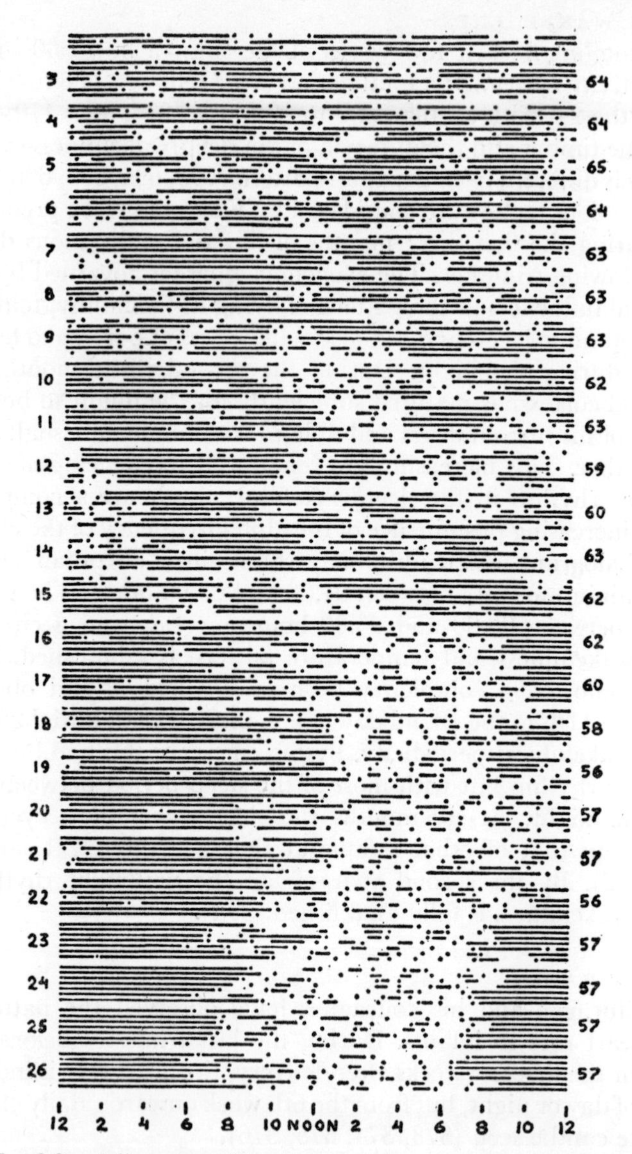

FIG. 8.2. Adjustment of the sleep/wakefulness cycle and feeding in an infant from the 11th to the 182nd day of life. The lines represent sleep periods, the breaks wakefulness, and the dots feedings. The weeks are indicated on the left, percentage of time spent asleep on the right. (Kleitman, N., Fig. 15.2, 470.)

sleep/wakefulness, is the only function to do so. The rhythm becomes more marked in the second and third weeks and resistance is notably higher during the morning. A nearly fully developed rhythm is seen between the first and fourth months.

Rutenfranz (781) has also detailed the age at which statistically significant day/night differences in other functions become evident: At the 1 per cent level of confidence, electrical skin resistance in the first week of life, wakefulness and body temperature in 2-3 weeks, urine excretion in 4-20 weeks (2-3 weeks at the 5 per cent level), heart rate, potassium and sodium excretion all at 4-20 weeks, phosphate excretion and creatine excretion at 16-20 months, and creatinine and chloride excretion (at the 5 per cent level) at 16-22 months.

EOSINOPHIL LEVELS AND ACTIVITY RHYTHM

The absence of morning eosinopenia in a group of infants less than 7 months old and its presence in most children in a group over 15 months has been noted by Halberg and Ulstrom (348), who suggest that this accords with the presence of a 24-hour rhythm in the older age group. Its absence in the younger group parallels the finding of Mullin, 1939 (654), who observed that by the age of seven months activity was mainly diurnal.

CORTICOSTEROID CONCENTRATIONS

Olivi and Genova (684) examined the plasma 17-OHCS level in babies from 3 to 12 months old and found a rhythm opposite in phase to the adult pattern in samples taken at 4-hourly intervals throughout the 24 hours with low values at 08.00 hours and maxima at 20.00 hours.

The development of the plasma corticosteroid rhythm towards the adult pattern has been elegantly demonstrated by Franks, 1967 (264). Comparing the 17-OHCS concentrations in a.m. and p.m. blood samples from groups of children, he showed that in premature infants 1-64 days old, the a.m. value was 14 per cent higher, in full term infants at 8-12 months it was 20 per cent lower, in children 22-26 months old, 38 per cent, and in children 3-13 years old, 79 per cent, as in the adult pattern.

TEMPERATURE

The first systematic attempt to study the development of the body temperature rhythm in infants was made by Jundell (443),

He obtained 4-hourly rectal temperature recordings over several days for one hundred children of up to five years. A rise in day temperature was seen by the end of the second month. No diurnal

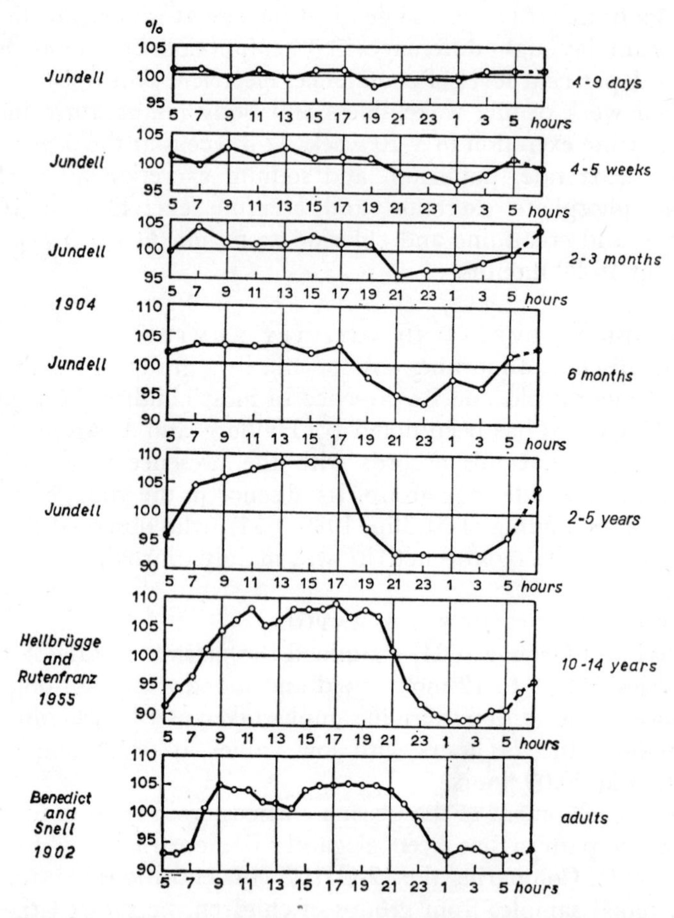

FIG. 8.3. The development of the circadian rhythm in body temperature in relation to age as shown by data of a number of observers. (Hellbrügge, T., Fig. 6, 372.)

variation in body temperature is seen during the first few weeks of life, but by six months the rhythm is clearly evident and during the second year of life it becomes fully established (469, 478 and see Fig. 8.3).

THE SYNCHRONIZATION AND MAINTENANCE OF HUMAN CIRCADIAN RHYTHMS

Synchronization of circadian rhythms with local time once they have developed, and their adjustment to the 24-hour cycle from an inborn periodicity of about 24·4 hours, depends on the presence of those environmental time cues which we term synchronizers or Zeitgeber. Light has been shown to be the dominant Zeitgeber in many animals, and appears to be very important in man also, but social factors may play a significant role.

ALTERNATION OF LIGHT AND DARKNESS

Some evidence for the part played by awareness of the alternation of light and darkness is seen in the sleep pattern of indigenous communities in Arctic regions in the perpetual summer daylight which is quite different from that observed during the winter darkness. Even subjects from temperate regions, when exposed to this continual darkness, tend to lose their normal sleep/wakefulness pattern though the ratio between number of hours of sleep and wakefulness is maintained (536). The excretory rhythms of indigenous subjects in these regions are also much less clear cut than those of people living in lower latitudes, suggesting that a normal daily alternation of light and darkness is necessary for the establishment and maintenance of circadian rhythms.

Lobban (536) has made a series of observations on two communities, one an Indian village at a latitude of 67° 40′ N and longitude 139° 40′ W and the other an Eskimo village latitude 70° 40′ N and longitude 160° 0′ W, and compared her findings with those on a group of subjects from Britain brought to the continuous daylight of the Arctic summer. Samples of urine were collected from each subject and the daily excretory patterns of water and electrolytes measured.

The rhythms of the indigenous subjects tended to be poorly defined, whereas those of the British subjects were maintained in their usual form. The Eskimos showed a particularly low amplitude in their excretory rhythms. These subjects are living at an even higher latitude than the Indians and are, therefore, deprived of a daily light/dark environment for an even larger part of the year. Diminutions in amplitude and other disturbances in the normal pattern of excretory rhythms have also been found in subjects from lower latitudes who have lived in the Arctic region for prolonged periods (533).

It might be objected that these poorly developed rhythms in indigenous Arctic dwellers reflect rather their habits than any direct effect of light. In the continuous daylight of the summer both Eskimos and Indians follow entirely irregular habits of life, nor do even members of the same family choose the same hours to sleep. This is not, however, an entirely adequate explanation, for in the Indian community studied by Lobban the habits of life in the winter were fairly regular, being determined by the school time of the children, although the excretory rhythms were even poorer than in midsummer.

This objection is also invalid for experiments in which subjects continued their customary activities, rising and breakfasting but remaining in darkness: the normal morning minima in eosinophil and lymphocyte levels were thereby delayed (735, 809). On the emergence of the subjects into the light a marked decrease in both types of white cell was seen. It thus appears that light plays a part in the control of the morning fall in eosinophil and lymphocyte numbers. It has similarly been shown (810) that the morning increase in urine flow can be postponed by keeping subjects in darkness for some time after waking. Urine flow recorded in a group of Spitsbergen miners showed little evidence of a circadian rhythm when the alternation of light and darkness returned in March (534). In blind persons also the eosinophil levels fall by a mean of 15 per cent between 06.30 and 09.30 hours as compared with a decrease of 38 per cent in controls (499). This is, however, a flattening and not absence of the normal morning eosinophil rhythm; and in some blind subjects normal renal rhythms may occur, though in others abnormalities have been demonstrated (540). Hollwich (399) has pointed out that in blind persons the sella turcica is relatively small and that the age of going blind influences its growth, the earlier the onset of blindness the smaller the sella turcica. He suggests that since the hypophysis cerebri is situated in the sella the smaller this structure the smaller the hypophysis is likely to be and that one may conclude that light may have a stimulating influence on the development of the diencephalohypophysial system. Migeon *et al.* (598) have, nonetheless, noted normal plasma corticosteroid rhythms in blind subjects. Although the amount of radiant energy reaching brain centres through the skull tissues may well be quite high (271) it seems reasonable to suppose that other Zeitgeber must operate in some or all of these persons.

Alternation of light and darkness with a period departing somewhat from 24 hours may be insufficient to entrain human rhythms (31, and see Chapter 7) and continuous exposure to high or low levels of illumination fails to influence human rhythms in the predictable manner observed in other species (922).

SOCIAL FACTORS AND OTHER POSSIBLE ZEITGEBER

Social factors have been suggested as important human Zeitgeber. Halberg *et. al* (334) have indicated that in eosinopenia, although light may play a role, social synchronizers are more important, and Aschoff (20) has suggested that maintenance of the body temperature rhythm may also be highly dependent on social Zeitgeber.

Four subjects living together in isolation may under some circumstances mutually entrain one another's rhythms to a common period, but under slightly different circumstances may fail to do so (725).

The persistence of circadian rhythms during experiments conducted underground would appear to exclude such influences as cosmic rays (682). A number of subtle geophysical synchronizers in animals and plants have indeed been suggested by Brown and his colleagues (113, 114, 115), though some of their findings have been criticised on statistical grounds (236).

DISSOCIATION OF RHYTHMS

Dissociation in circadian rhythms may occur or be produced under a number of circumstances such as the imposition of abnormal time routines, rapid phase shifting across a number of time zones or conditions of complete isolation from the normal human Zeitgeber.

In a series of experiments in the Arctic Lobban and her colleagues (513, 514, 515, 517, 533, 535) have shown that in subjects attempting to live on a 21, 22 or 27-hour 'day', sodium, chloride and water excretions were usually closely connected, while potassium excretion often became dissociated. 24-hour periodicity was particularly stable in potassium excretion, persisting in one subject after five weeks on 22-hour time and in another subject after six weeks on 27-hour time while other rhythms had adopted the artificial day length.

After flights across a number of time zones temperature rhythms appear to adapt in a matter of days whereas adrenocorticosteroid excretion may take some weeks (164, 251).

ENDOGENOUS CIRCADIAN CLOCK

Synchrony between the sleep/wakefulness cycle and the excretory rhythm of potassium and, for several weeks, that of chloride in a subject under conditions of prolonged isolation, as one of us has shown (607), suggests that there is some intrinsic rhythmic process—a biological 'clock'—governing circadian rhythmicity. The occurrence of free-running rhythms departing slightly from a cycle length of exactly 24 hours has now been observed on a number of occasions in man (29, 30, 32, 153, 346, 607, 609, 818, 819). This departure from a precise 24-hour period is strong evidence for the existence of an endogenous circadian clock.

The dissociation between various circadian rhythms raises the further possibility that there may be more than one clock. This is, however, not a necessary inference. As soon as any routine such as abnormal time schedule begins, the body is subjected to two conflicting rhythmic influences, the endogenous circadian clock and the new routine of daily habits. Competition between the two influences has, indeed, been shown by the occurrence of 'beats' (515) and it would seem that certain variables, such as excretion of sodium chloride, are much more susceptible to environmental influences than others, such as potassium excretion. To prove the presence of more than one circadian clock, it would be necessary to demonstrate the existence of two rhythms each with a slightly different cycle length, and both endogenous.

The most convincing account of such dissociation (30) is of a subject who, living alone in complete isolation from time clues, followed an activity cycle of 33·2 hours while his temperature rhythm showed a period of 24·8 hours. It is stated that one other subject showed a similar desynchronization throughout his period of confinement, and seven others, out of 50 subjects studied, showed a similar desynchronization after they had been confined for 9 to 23 days. The only weakness of this argument lies in the assumption that one's habits of activity and sleep necessarily reflect an endogenous rhythm of alertness and sleepiness in all circumstances. In a function amenable to some degree of voluntary control, and susceptible to a variety of psychological influences (see Chapter 7), such an assumption cannot be made with entire confidence.

RHYTHMICITY AT VARIOUS LEVELS

There is, however, no inherent difficulty in supposing that circadian rhythmicity resides in many cells and organs of the body, and that the action of the 'transmitter' mechanisms, such as the hormones (Chapter 3), is to synchronize their rhythms rather than to impose a rhythm on a non-rhythmic tissue. Rhythmicity may, indeed, exist at many levels including the cellular. Tharp and Folk (863) have demonstrated how rhythmic changes in mammalian heart cells may persist in isolation and clocks exist in uni-cellular organisms (358) and even in a single metazoan neurone (850). A rhythm at one level may also be controlled by a rhythm at another level (Fig. 8.4). For example, the eosinophil rhythm is controlled by circadian variation in the corticosteroid rhythm, which in turn is controlled by the ACTH rhythm, itself moderated by CRF, presumably governed by a higher centre in the brain, which in its turn is influenced by sensory input from the external environment such as awareness of the alternation of day and night.

POSSIBLE UNDERLYING MECHANISMS, OSCILLATORY AND ELECTRONIC MODELS AND THE CHRONON CONCEPT

Hendricks (380) has suggested that the possible controls involved in circadian systems may be of four types, which he has designated centre point regulation, product action on input into a reaction series, desuppression of enzyme synthesis at the level of the gene, and multiple hormone action in a series, with eventual feedback. Any or all of these could be involved in man but genetic control and the hormonal mechanism appear particularly relevant (380, 389).

Temperature independence of the biological clock is a marked feature which Pittendrigh and his colleagues (720, 721, 722, 723) have suggested may be explained by the existence of two coupled oscillators. One oscillator, the A oscillator, is taken to be relatively temperature-independent, and another, the B oscillator, is supposed to be, in part, temperature-dependent. The A oscillator is imagined to be capable of adjusting its phase immediately in response to a new Zeitgeber, even perhaps to a single signal. For

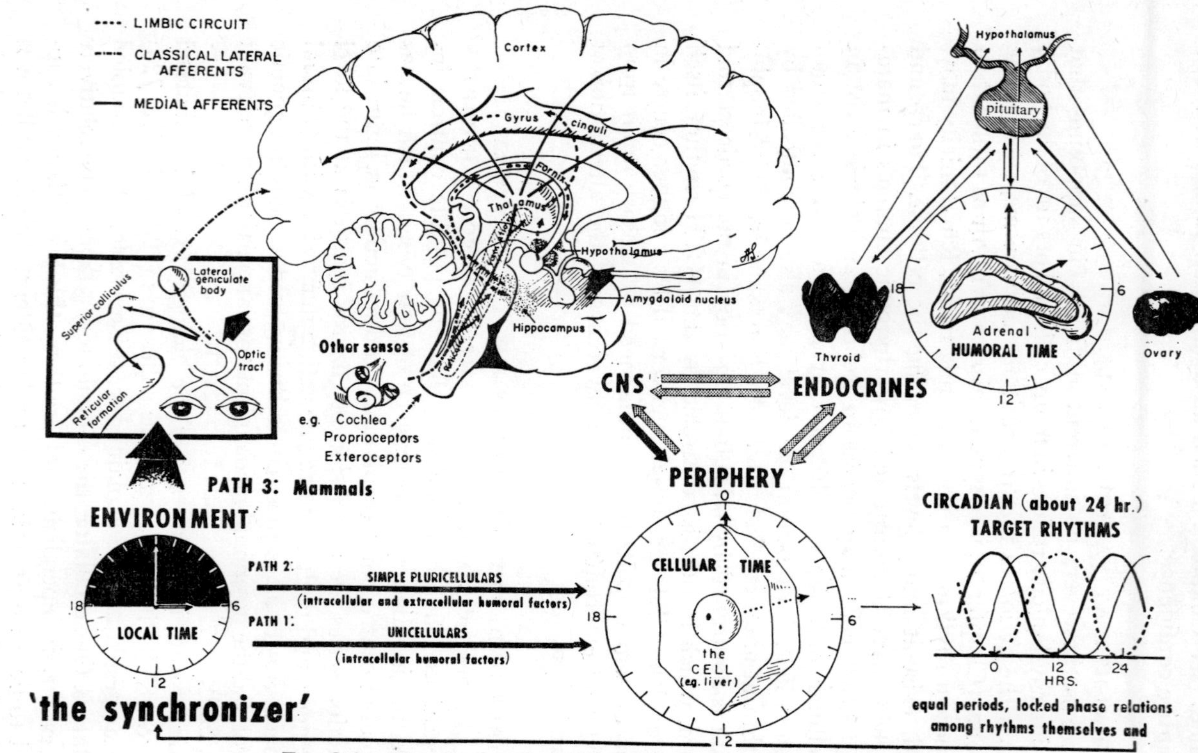

FIG. 8.4. Sketch of pathways and factors which may be involved in the synchronization of circadian rhythms with one another and with the external environment. (Halberg, F., Fig. 8, 323.)

a time the B oscillator resists the new phase of A but ultimately becomes driven by it. The model is able to account for the response to phase shifting in mammals and may well be applicable in man. The phase of the human clock can be reversed by changes of habit or environment after a delay whilst the Zeitgeber are bringing it into synchrony with the new schedule. Changes of cycle length, on the other hand, are very difficult to induce.

It has recently been claimed that the operation of the clock can be influenced in man by a weak alternating rectangular electrical field (920). Without their knowledge subjects were exposed for a part of a period of solitary confinement to such a field; while they were thus exposed, the period of their activity and temperature rhythms shortened though it still exceeded 24 hours, and they spent a larger proportion of their time active and a smaller proportion at rest; the pattern of their temperature rhythm also changed in such a way that the falling phase became slow and the rising phase rapid, and the temperature maximum then fell earlier in their activity period. It is not easy to see what sensitive structure, nervous or other, in the human body can detect or respond to such an electrical field, so the full interpretation of these observations is, for the present, obscure.

Other oscillator models of circadian mechanisms have been formulated (138, 222, 298, 299, 358, 360, 450, 645, 802, 837, 850), as have mathematical models based on electronics (222, 800, 801, 917, 918, 919). Recently Ehret and Trucco (223) have put forward a suggestion, which they term the chronon concept, that circadian rhythms are based on the recycling of a mechanism regulating the transcription of template RNA from DNA (Fig. 8.5). The sequence of steps in this process they refer to as the sequential transcription (ST) component of the circadian clock escapement. This sequential transcription component is considered to be a very long polycistron, i.e. complex of DNA, which they name the chronon and define as a set of cistrons and codons. The chronon alone can allow for the escapement of a 'day' clock, a chronon recycling (CR) component is invoked to provide for the reiteration of each day programme. The transcription rate is taken to be influenced by regulators which are relatively temperature independent.

Iberall and Cardon (416), in a very interesting physical review of regulation and control of biological systems, suggest that there is a common circadian oscillator and that its sensitivity indicates a general Van der Pol relaxation oscillator (885) rather than a continuous sine wave resonator of a precision clock.

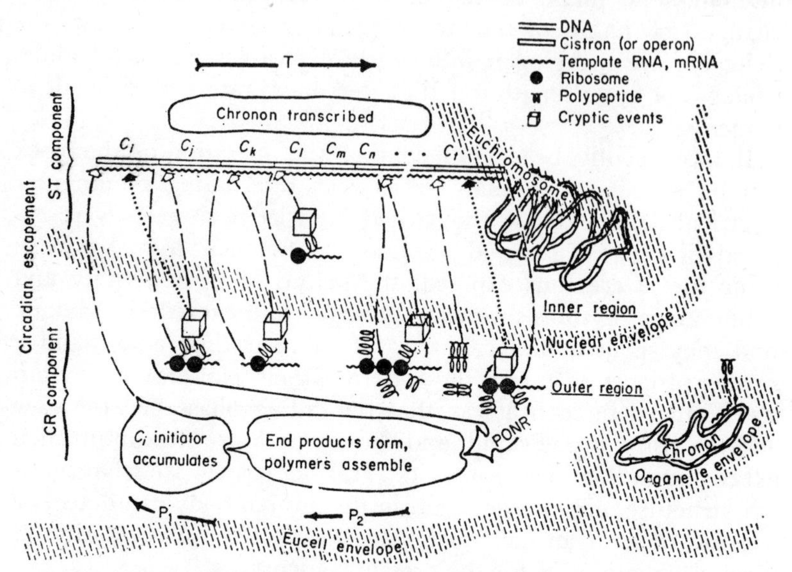

FIG. 8.5. Schematic representation of the chronon concept. In this the circadian cycle is taken to consist of pretranscriptional (P_1), transcriptional (T) and post-transcriptional (P_2) phases. The chronon is the sequential transcriptional (ST) component, one of the many very long DNA polycistron complexes on a single eukaryotic chromosome or in a cell organelle. Transcription of template RNA proceeds to the right, beginning at the initiator cistron (C_1). This part of the cycle is slow and relatively temperature independent. Protein synthesis continues after transcription of the terminator cistron (CT) and an initiator substance (C_1 initiator) accumulates. This part of the cycle (P_2, P_1) is very temperature dependent and constitutes the Chronon Recycling (CR) component. The wide white arrows indicate 'transcription starts', the wide black arrows 'transcription stops'. Diffusion paths of the effector macromolecules are shown by broken and dotted lines.
(Ehret, C. F. and Trucco, E., Fig. 1, 223.)

The location of the human biological clock cannot yet be decided with any certainty. The cerebral cortex can be excluded since some rhythms persist even in the absence of cortical function. The hypothalamus may tentatively be suggested as a most likely site.

Chapter 9

Experiments with Abnormal Time Schedules

Non-24-hour Days—Phase-Shifts—Time Zone Transitions.

We can attempt to avoid the innumerable periodic influences of nychthemeral existence by escaping from time. Such attempts, by living in natural caves or in artificial underground chambers, have already been described. This chapter will be devoted to the alternative method of study, the deliberate manipulation of time schedules. Some repetition of results already described in other chapters is inevitable; but since most studies on abnormal time schedules have involved several different physiological functions, they are here assembled together.

It perhaps needs emphasis that the most interesting outcome of such experiments is the frequent persistence of a 24-hour rhythm which does not correspond to the subject's habits. When a man lives, for instance, on a 21-hour day, then any aspect of his habits, his mealtimes, waking and sleeping, rest and exercise, can impart a 21-hour rhythm to the function being studied, without the clock itself becoming adapted or entrained to the new time schedule. If, however, 24-hour influences have been rigidly excluded, then any persistence of 24-hour rhythmicity must result from the operation of the clock. When over the course of days, weeks or months a 24-hour rhythm fades out and is superseded by a 21-hour rhythm, this may only indicate a weakening of the influence of the 24-hour clock, rather than, as is sometimes implied, its adaptation to a new period. Much the same argument applies when the phase of environment and habit is abruptly changed.

NON-24-HOUR DAYS

The assumption of a 12-hour cycle of activity has a well-defined but limited usefulness. It provides a simple alternative to keeping a man under conditions of absolute constancy, which is always attended by the suspicion that quite minor rhythmic influences assume a great importance when the major influences have been removed. On the 12-hour schedule the major influences of sleep, activity and meals continue, but are so contrived that every hour in the 24 can be matched against another, 12 hours earlier and later, in which all these rhythmic influences are the same. Comparison between any pair of hours, such as midday and midnight, will thus reveal the operation of the clock upon an identical

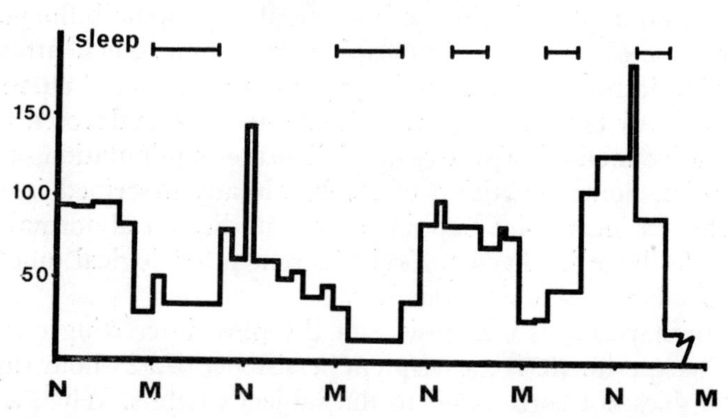

Fig. 9.1. Potassium excretion, μEq/min, by a subject sleeping at first for 8 hours in every 24, then for 4 in every 12. N, noon; M, midnight

background influence, and inherent circadian rhythms can be demonstrated in a conveniently short period of study. By this technique, the endogenous rhythm of body temperature (470, 603, 613), of urine flow and pH and excretion of sodium, potassium (Fig. 9.1) and chloride (613), and of eosinophil count (338) have been established, whereas alveolar tension of carbon dioxide, and renal excretion of phosphate, were shown to have little if any endogenous rhythm (613); phosphate excretion on this schedule immediately assumed a 12-hour periodicity, with little sign of a 24-hour influence. Even after a long time on this schedule, or on a 48-hour 'day', there is no sign of a waning influence of the 24-hour clock (338, 470, 579) on those functions whose variation remains circadian. It may be supposed that on both these schedules the clock receives regular reinforcement from the habit rhythm, so they are not very suitable for studying the persistent influence of the clock over long spells of time.

One study in seeming contradiction is the claim (521, 686) that after only a few days on a 12-hour schedule the concentration of cortisol in the plasma may adopt a 12-hour periodicity. The experiment was designed to demonstrate the direct influence of sleeping habits, rather than to search for any circadian influence, so subjects adopted three different sleeping schedules, settling down for 4 hours of sleep at 00.00 and 12.00, at 06.00 and 18.00, and at 10.00 and 22.00 hours. The individual results are somewhat irregular; the authors present also the means related to the subjects' hours of sleep: values during sleep were averaged, then those in the hour immediately after waking, the next hour, and so on. In this way, the direct effect of the sleeping routine becomes apparent, and any circadian influence is eliminated. It would be interesting to see also the means at each hour of day, so that any persistent circadian influence could likewise be assessed. These experiments do, however, suggest a stronger influence of habit than has been found by other workers.

In the studies thus far reported, the subjects were living in communities who were following usual nychthemeral habits. This is more difficult on a 'day' differing slightly from 24 hours, since subjects will repeatedly drift out of phase with people around them, and back into phase with them; since one object of such experiments is to study the long-term persistence of the influence of the endogenous clock, it is desirable to eliminate such social influences, as well as the accompanying alternation of light and darkness. The continuous daylight of an Arctic summer offers a laboratory congenial to the tastes of many, where the only obvious 24-hour periodicity is the passage of the sun around the horizon, and even this may be obscured by cloud for weeks on end. The tides, and the consequent feeding habits of birds, convey an obvious rhythm, but one which departs notably from a 24-hour period. An alternative situation, from which periodic influences are more rigidly excluded, is a deep cave or underground laboratory.

Apart from the persistence of individual components of rhythmicity, dissociation between different components can provide useful clues to their causal connexions. If a particular variable loses its 24-hour rhythmicity, it is hardly likely to be the immediate cause of the persistent 24-hour rhythm of another variable.

Some slight information has been obtained even from subjects living on days of 21 and 28 hours in a nychthemeral society (470); of two subjects attempting this, one had a temperature rhythm which entirely failed to adapt, and there is some suggestion that his sleep/wakefulness rhythm was also unadapted.

More satisfactory conditions were attained at Tromsö (475), where the communal life of a family of four living on an 18- and 28-hour day would, it was hoped, conduce to good adaptation; there were, however, persistent 24-hour influences, despite the continuous daylight, in the form of considerable temperature swings, and the very regular nychthemeral habits of the local community; the subjects comment: 'It was very depressing to have to walk through deserted streets, with the stores all closed, the broad daylight only accentuating the ghost-town aspect of the surroundings'.

These continuing 24-hour influences do not detract from the value of this experiment, since some interesting dissociations between rhythms emerged. In all three subjects (one was excluded through illness) pulse rate promptly assumed an 18- or 28-hour rhythm; but temperature, on both schedules, adhered obstinately to 24-hour time in one subject in whom soundness of sleep, as assessed by diminution in motility, also failed to adapt to the artificial schedule. It thus appears doubtful whether there is any endogenous rhythm in pulse rate, and the endogenous rhythms in sleep and in temperature seem to be connected.

The same conclusion emerged even more clearly from a study on two subjects spending 32 days in the Mammoth Cave of Kentucky, living on a 28-hour day; one adjusted both his sleep and temperature rhythms easily, the other failed to adjust either. The temperature of the cave was absolutely uniform, and neither daylight, sound nor other nychthemeral clue penetrated. A further point of great interest emerged: the subject whose rhythms adapted easily to the 28-hour schedule had some difficulty in readapting his sleeping habits to a 24-hour day when he emerged. The only influence which could have imposed such a 28-hour period, when he was living again in a nychthemeral society, was an entrained clock, and this is one of the clearest pieces of evidence yet available for entrainment of the clock itself. His temperature rhythm reverted immediately to a 24-hour period, but was of low amplitude for some time, suggesting that a temperature rhythm of normal amplitude needs something beyond the direct influence of nychthemeral habits (470).

In another study six quadriplegic patients, who were entirely dependent upon the nursing staff for their movement and posture, as well as for intake of food and drink, were subjected to a 19-hour schedule (523), and their renal function was studied. Excretion of sodium, potassium and chloride all showed a persistence of 24-hour rhythmicity with an added period of 19 hours, least evident for potassium; by contrast, urine flow and urea excretion seemed to be largely determined by periodic intake, and thus showed a 19-hour rhythm with some dissociation from electrolyte excretion.

The most extensive series of experiments upon days of abnormal length are those carried out by Lobban and her associates during three separate summers in Spitsbergen, in the course of which a large number of functions were measured, providing evidence for association and dissociation as well as for the varied persistence of different functions, and for the differences in behaviour of different individuals. Temperature as well as daylight were here fairly uniform over the 24 hours; but subjects engaged in activities which at times interfered with the physiological observations. On one expedition, for instance, systematic meteorological observations required the subjects to be aware of normal as well as of the artificial experimental time. In the first expedition (513, 517), after a preliminary period to define their normal rhythms, the eight subjects lived on a day of 22 hours, secured by using watches so adjusted that they recorded 12 hours when only eleven had elapsed. The persistence of a 24-hour rhythm in renal behaviour was demonstrated even after living for five weeks on a 22-hour day. This was most conveniently shown by observing a continuous period of 12 days on experimental time, corresponding to eleven real days. If at the start of this spell real and artificial time were coincident, then the amplitude of the excretory rhythms, alike for sodium, potassium, and water, was fairly large. After five or six days, artificial time was exactly out of phase with real time and the amplitude of the rhythms was very low, but it regained its previous size after a total of twelve experimental days. The numerous influences arising from the subjects' meal time, sleep, activity and so forth following a 22-hour cycle were thus obviously competing with a persistent 24-hour influence. Some subjects adapted better than did others; and, though there was no clear dissociation between different components of the rhythm, the 24-hour influence seemed stronger for potassium than for sodium or water excretion.

In a second series of experiments (514, 515) these workers adopted a 21- and a 27-hour day for two groups each of six subjects. If both times were initially in phase at midnight, they would again be in phase after 7 or 9 days by real time (8 days by experimental time). It now became clear that there were differences between the adaptability of one subject and another. When

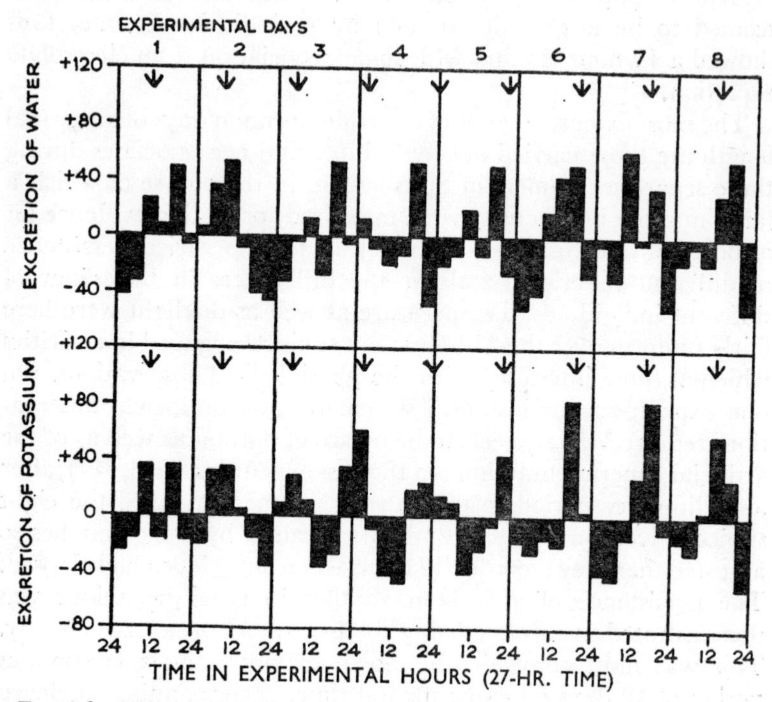

FIG. 9.2. Water and potassium excretion, percentage deviation from mean, in a subject during eight 27-hour 'days'. Arrows indicate midday by solar time. (From Lewis, P. R. and Lobban, M. C., 515.)

living on 21-hour time over a period of a real week a non-adapter would show seven peaks of potassium excretion whilst a perfect adapter would show eight peaks. Similarly with subjects on 27-hour time, within eight experimental days a non-adapter would show eight excretory peaks (Fig. 9.2, upper) whilst a perfect adapter would show nine corresponding to the number of real days elapsing (Fig. 9.2, lower). Different subjects, and different constituents in the same subjects, showed either form of behaviour, but when experimental and real time were exactly out of phase the amplitudes of the rhythms tended in all subjects to be low, suggesting that the 24-hour rhythm and the artificial rhythm were both present and competing with one another (Fig. 9.3). While some subjects adapted to the artificial time promptly others adapted slowly over some weeks and in one subject the 24-hour rhythm of potassium excretion was completely unadapted even after 6 weeks on 27-hour time. A criticism which has been levelled against these experiments is that the apparently persisting 24-hour clock must have been remarkably accurate. By dead reckoning it was supposed that after each week experimental time and real time should be in

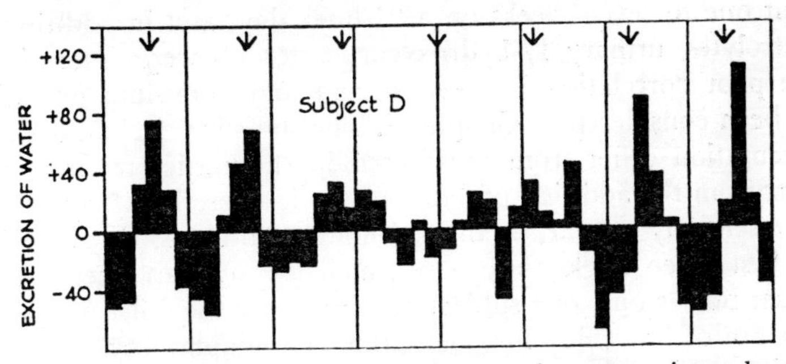

FIG. 9.3. Urine flow, percentage deviation from mean, in a subject during eight 21-hour 'days'. Arrows indicate midday by real time. (From Lewis, P. R. and Lobban, M.C., 515.)

phase, and the best non-adapters appeared indeed to be perfectly in phase after 36 real days. If one accepts a drift of 3 hours as the maximum that could escape detection, this represents a drift of not more than 5 minutes per day in comparison with an error of 30-60 minutes which the human circadian clock has been supposed on other grounds to show (see Chapter 7). However, Aschoff and Wever (32) in free running experiments of an entirely different type found mean cycle lengths for some subjects very close to 24 hours; and since only two of the twelve subjects of Lewis and Lobban showed so accurate a clock this is perhaps not beyond the bounds of credibility. It does, however, suggest the presence of an external 24-hour Zeitgeber (25).

Subjective accounts of the ease with which subjects slept indicated that at least two had difficulty in sleeping when real and experimental time were out of phase (101, 517). The data do not, however, permit a critical study of the association between wakefulness and other rhythmic functions. Of other functions studied, temperature rhythm usually adapted rapidly and completely to artificial time and thus became completely dissociated from renal rhythms. Of the renal rhythms potassium was the most persistent and thus sometimes became dissociated from water, chloride and sodium, which were following experimental time while potassium followed 24-hour time. Since excretion of sodium and of water are much more liable to influence from the varied activities of the day than is potassium excretion, it is not surprising that they escape sooner from the influence of a circadian clock.

In a third expedition to Spitsbergen (826), seven subjects spent one to seven weeks on a 21-hour day, and in addition to electrolytes, urinary 17-hydroxycorticosteroids were determined. The poor correlation between steroid and potassium excretion has been considered in Chapter 4. The statistical techniques and presentation differ from those used in the earlier reports, but, apart from the addition of data on steroid excretion, the findings are essentially similar, in that 24-hour rhythms persisted during the first three weeks, but were much less obvious after 5 or 7 weeks; but as only one subject continued for the full 7 weeks the data are of less value for comparing the adaptability of one or another subject than are those from the earlier expeditions. One new observation of particular value was made: five of the subjects were studied immediately on return to 24-hour time, when all the excretory rhythms immediately resumed a 24-hour periodicity, without trace of a 21-hour influence. One of these was the subject whose rhythms had become well adapted to 21-hour time after 7 weeks on this schedule. We must conclude, at variance with Kleitman's observations upon sleep rhythms (470), that the clock itself had not become entrained to the abnormal time, but that such influence as it continued to exert was overruled by the direct influence of 21-hour habits. As has been repeatedly emphasized, a 21-hour rhythm in a subject following a 21-hour routine is no evidence for the operation of any form of 21-hour clock.

It is difficult to reconcile the steroid findings in Spitsbergen with the claim (686) that plasma steroid rhythms can be rapidly entrained to a day of 19 to 33 hours.

In bunker experiments on days of abnormal length (31) two subjects adapted their sleeping habits to an artificial long day although their temperature rhythm was not entrained, and in one subject who adopted sleeping habits to conform with an artificial short day the temperature rhythm was likewise not entrained. It is difficult to be certain, however, whether these subjects merely adjusted their sleeping habits out of convenience to accord with the day length imposed by the illumination of the chamber, without any entrainment of their sleep/wakefulness cycle. These observations do not, therefore, conflict with the supposition that the endogenous rhythms in temperature and sleepiness are closely and perhaps causally connected. In three other subjects who adjusted their sleeping habits to the artificial long day the temperature rhythm was also entrained to this period.

In an essentially similar experiment (300, 426) two subjects spent 11 real days living on a 22-hour day. Their mealtimes and sleeping times were deliberately adjusted to the artificial short day, of whose duration they were unaware, and their urinary excretion patterns were analysed for conformity to a 22-hour and a 24-hour period. The existence of both periods was demonstrated, although the amplitude of the 22-hour rhythm was greater for almost every constituent in both subjects. A point of special interest is that the phase relationships for the two periods were different; the 24-hour rhythm consisted of an excretory maximum at or shortly after midday for flow and sodium, chloride and potassium, and slightly earlier for calcium and magnesium, while the maximum for the 22-hour rhythm was somewhat later, thus confirming the supposition (see Chapter 4) that the direct influence of meals and other habits is to impose a rhythm with an excretory maximum somewhat later than that due to the endogenous rhythm. The special interest of this experiment was that one of the subjects was a psychiatric patient who suffered from a very regular 48-hour oscillation in mood. This mood rhythm immediately and completely adapted to the new length of day, and was thus dissociated from the urinary rhythms.

PHASE-SHIFTS

As with artificial alteration of day length, a sudden phase-shift may be achieved by altering the clock and adjusting habits to conform with it. Unlike alteration of day length, however, this is a change often made for other than experimental purposes. The commonest of such adjustments, night work, offers rather unsatisfactory material for physiological study upon rhythmicity, since the time-shift is hardly ever perfect; the time of work may be adjusted, but the worker lives in a nychthemeral environment, he commonly adjusts his times of sleep and leisure to the convenience of his family or of society at large, and he usually reverts to nychthemeral habits for a long week-end. Many observations upon night workers are concerned rather with the practical issues of their performance at work, satisfaction with the arrangement, incidence of sickness and so forth, and will be considered in Chapter 10.

The other common form of phase-shift occurs on flight into a different time-zone. Here all aspects of environment suffer the same phase-shift: the subject must abruptly change his notion of time with regard to the hours of light and darkness, the habits of people around him, and the time he is expected to sleep, eat, and work. Ideal conditions are thus presented for scientific study but air travellers are commonly too fully occupied with the purpose of their journey to be available for the necessary examination. In this chapter we will consider data derived from artificial phase-shifts achieved, for example, by adjustment of time in an Arctic summer, and from air travellers who have been able and willing to participate in systematic studies. Findings relevant rather to the practical implications of long flights will be considered in Chapter 10.

In one of the earlier attempts to impose artificial phase-shifts (465), titratable acid excretion appeared to remain higher by night than by day in subjects who reversed their habits, but the urine was divided into two aliquots, of 8 hours for the sleeping and 16 hours for the waking period. Some other similar studies are mentioned in Chapter 4. A more substantial study on phase-shifts is reported by Völker (893), who kept subjects recumbent in a darkened room in which lights were on only during their waking hours, and made observations upon a normal routine with sleep from 23.00 to 08.00 hours, and on routines in which sleep was 8 hours earlier, 7 hours later, and shifted by 12 hours. In some experiments the subjects took meals at customary hours, suitably adjusted to their altered hours of sleep, and in some the subjects fasted. The different experiments only extended over 2 to 4 days, so their value is rather in demonstrating the endogenous nature of the rhythms which are not immediately dependent upon habit, than in assessing the possibility of entrainment to a new time, for which they were much too short.

He records that subjects had difficulty in sleeping at unaccustomed times, often lying awake or sleeping lightly, whilst they often found it hard to keep awake between 03.00 and 06.00 hours. He also measured temperature, pulse rate, blood pressure, urine flow and nitrogen excretion, and ventilation and oxygen consumption. These very extensive data deserve a more rigorous treatment than they have previously received; since temperature and pulse rate were recorded hourly in most experiments, the published means have been subjected to Fourier analysis, with the results shown in Table 1. The fasting experiments were performed upon a single subject, while four participated in those in which meals were taken. It will be seen that the amplitude of the pulse and temperature rhythms was little altered, and that the phase was at most partially shifted in the appropriate direction.

TABLE 1

FOURIER ANALYSIS OF DATA OF VÖLKER, H. (893)

		Control	Sleeping 8 hours earlier	Sleeping 7 hours later	Phase reversal
Pulse,	Amplitude	6·8	4·1	4·4	...
beats/min.	Acrophase	15.30	11.00	19.12	...
Temperature °C	Amplitude	0·41	0·26	0·25	...
	Acrophase	16.16	13.45	17.42	...
Temperature °C	Amplitude	0·29	0·18	0·31	0·23
fasting	Acrophase	16.15	14.38	20.02	16.12

Since these experiments only lasted for a few days, and circadian rhythms persisted in all variables studied, they are of no use for assessing the connexion between separate variables. The same is true of another series (603) in which a group of subjects spent 6 days in Scotland in midsummer living on a reversed routine, and failed to reverse their rhythms of sleepiness, of temperature, or of urine flow. A much more extensive series of experiments has been carried out by Sharp and his colleagues in the Arctic, in which they reversed their routine, using darkened tents to simulate night, or wearing blindfolds. In an earlier expedition with 6 subjects (807, 808), they measured urine flow, pH and specific gravity, renal excretion of sodium and potassium, and counts of the different types of leucocyte. After 6 days all these variables had completely reversed their rhythm to conform with the new sleeping habits except in one subject, who was slow in reversing his excretory rhythms, as also in sleeping at unfamiliar times, although all circumstances except an intrinsic rhythm should have been conducive to sound sleep.

In a second expedition (570, 813) urinary ketosteroids, ketogenic steroids and creatinine were also measured, and the four subjects were studied every second day for 8 days after reversal of habits; after a fortnight on reversed habits they were again observed every second day for 8 days after the return to normal time. In the second day after reversal of habits there was obvious evidence of conflict between the persistence of the former rhythm and the rhythm of the actual habits. The sodium and potassium excretion were still lowest during the sleeping period, but they were not as low as were the previous night's excretion rates and the excretory patterns during the day were considerably disturbed. After 4 or 6 days the pattern was very similar to that before reversal, and complete adaptation had occurred. There was, however, clear dissociation between the adaptation of different functions; excretion of ketosteroids was fully adapted even on the second day, and seems therefore to be wholly dependent upon habit; sodium and water excretion were fully adapted after 4 days, but potassium and ketogenic steroids took 6 or 8 days to become fully adapted. The ketogenic steroids are the characteristic product of the adrenal cortex, so these observations lend some slight support to the idea that the adrenal rhythm is involved in the rhythm of potassium rather than of sodium excretion. These data were not subjected to any formal analysis, and inspection suggests that many of them could not be well fitted to a sine curve; but the high outputs during the period of sleep when the subjects were imperfectly adapted, and the gradual shift of the mode, suggest that partial adaptation was manifested both by low amplitude and by incomplete phase shift. It is clear, however, by comparison with the observations of Lobban and her colleagues, that circadian rhythms in general can adapt more readily to a phase-shift than to an alteration in cycle length.

When after a fortnight the subjects reverted to normal time, the pattern of return to normal of the different excretory rhythms was remarkably similar to that of the initial adaptation to a nocturnal existence. This observation is particularly important in view of the theory (112) that the postulated circadian clock is in fact a response to some undetected geophysical rhythm. If this were so then one might expect a slower adaptation to abnormal time, and the readaptation to normal time should be instantaneous. The closely similar course of readaptation of all functions suggests that the controlling clock had itself become adjusted to the phase-shift, and contrasts strikingly with its apparent inability to adapt to a different cycle length even after many weeks. A similarly rapid but incomplete phase change in renal electrolyte rhythms has been found (281) in subjects remaining recumbent in bed but exposed to a phase reversal in lighting.

TIME ZONE TRANSITIONS

An abrupt shift in the phase of external rhythms is achieved every time anyone flies across several time zones. Here no artificial contrivance is necessary to ensure complete phase-shifting of all external rhythms, both climatic and social, but considering how often people do thus travel the data on their physiological responses is still sparse. Some observations confined to one or a limited range of variables will first be considered, followed by a few studies which cover a wider range. Temperature rhythm in one study (133) was found to take three to four days to adjust to local time after a 5-hour phase-shift from Canada to England. In another study (789), on a flight from Tokyo to Kansas, a 10-hour shift, it was again noted that a few days elapsed before entrainment to the new time, and further that a 6-year-old child adjusted more readily than two adults.

Pulse rate, as has been seen in Chapter 5, shows very little evidence of an endogenous rhythm; but continuous records on four subjects during transatlantic flights (406) have demonstrated a slow pulse at times when the subjects should, according to their usual nychthemeral habits, have been asleep.

Renal rhythms have been observed (280) in two subjects flown from Amsterdam to New York and back on two occasions. In the first experiment they remained for 4 days in New York before returning to Holland, in the second they returned after a stay of only 2½ hours at the New York airport. In the first experiment disturbances of sodium, chloride and potassium occurred on transfer to New York. These took the form of diminution in the amplitude of excretion and lack of synchrony with local time. In 4 days they were still not fully adapted to the new environment, but on returning to Amsterdam they readapted to local time within a day. In the second experiment some slight diminution of amplitude of urinary volume and electrolyte excretion occurred after the return flight, but a normal pattern was re-established the next day.

Essentially similar observations have been made on trans-polar flights between Paris and Anchorage in Alaska, involving a time shift of 11 hours in about 11 hours flying time (491, 492). Four-hourly urine sampling for 7 days in Paris revealed that of the various constituents analysed, potassium and 17-hydroxycorticosteroids showed the largest and most consistent circadian rhythms; these were therefore chosen for further study, and a mean curve was calculated for the 7 control days on 8 subjects. All eight were then flown to Anchorage, and back after a stay of 20 hours, from 02.15 to 22.00 hours by Paris and from 15.15 to 11.00 by Anchorage time. Their mean excretory rhythms for both potassium and steroids were hardly altered.

In a second experiment, two of the subjects flew again to Anchorage and stayed there for 5 full days (140 hours) continuing to collect urine in 4-hourly aliquots. The rhythm in excretion of potassium and steroids was somewhat disorganized on the days of flight in both subjects; one then reverted to Paris timing in both these urinary rhythms, only adapting to Anchorage time after about 4 days; in the other subject the rhythms never adhered to Paris time, and had assumed Anchorage time after 2 days. The time course of adaptation of steroid and potassium excretion was roughly similar, and only an occasional single urine sample showed obvious dissociation between them. Partial adaptation was shown both by a partial shift of acrophase, and by the appearance of two separate peaks of excretion; only rarely was there any flattening of the rhythm.

A comparison has been made between the effects of a 10-hour time shift after an east-west flight from Oklahoma to Manila, a 7-hour shift in a west-east flight from Oklahoma to Rome, and a north-south flight of similar duration from Washington to Santiago (362, 363, 364). Observations were made for a week before flight, for 8 to 12 days at the destination, and for a week after return. The circadian rhythms in rectal temperature and pulse rate took around 4 days to adapt to the new time zone, and a comparable time to re-adapt on return; palmar evaporative water loss adapted rather more slowly. None of these rhythms was disturbed by the north-south flight. Fatigue assessed subjectively, and differential reaction time, were increased similarly by all three flights, but only for the first day or so; the authors did not, however, record circadian variations in tiredness, which is the aspect of practical importance; it is no bad thing to be tired when one plans to sleep. Moreover, the test used (700) appears to be somewhat insensitive at detecting persistent rhythms of tiredness after a transatlantic flight (259). It is thus not clear from these papers how far failure of adaptation of physiological rhythms was associated with any measurable aspect of deterioration in performance. The authors note a number of minor differences between the consequences of the east-west and the west-east flights, but several factors other than the difference in compass direction may account for these. Only two subjects took part in both flights, and, as in the observations of other workers, some subjects seemed to adapt sooner than others.

One of the earlier reports, which includes a variety of functions, is by Flink and Doe (251) who observed a single subject flying from Minneapolis to Japan, a 9-hour time shift, and later to Korea (9½ hours). Subjective symptoms included extreme sleepiness between 15.00 and 22.00 hours (local time) on the day of arrival in Japan despite adequate sleep on the plane *en route* and wakefulness between midnight and 06.00 hours (local time). This disruption of the sleep/wakefulness pattern persisted to a gradually lessening degree for about 3 weeks. Changes in urinary corticosteroids were also studied in this subject. It was found that the timing of his corticosteroid excretion became synchronised to local time nine days after leaving Minneapolis, but that in two months normal amplitudes had not been achieved.

Potassium and sodium excretion were also adjusted to the new time after 9 days, although their pattern was not quite like that observed in America. When the detailed pattern of a rhythm is somewhat altered even some months after a time shift, it is always possible that this is a permanent state and reflects a somewhat different pattern of life. The major change in this subject was apparently complete in just over a week, and there was no obvious dissociation between the different components such as would throw light upon their causal relationships.

A brief report has appeared (327) of temperature, pulse rate, and time estimation on a single subject who, after a month of observation in the United States, flew to Japan and returned 10 days later for a final 6 weeks of observation. The temperature and pulse rhythms had their acrophase around 18.00 hours, and counting was slowest around 06.00 hours, by American time. The phase took about 4 days to adapt after the time-zone transition; the acrophase for pulse and temperature were roughly coincident throughout but the rhythm in time assessment seemed to adapt rather more slowly after the return flight.

We ourselves with our colleagues in Manchester have made a number of similar studies on the effects of time-zone shifts in air travellers, which are as yet published only in abstract form (161, 164, 229, 259).

One concerned a family consisting of husband, wife and four-year-old son whom we observed before a flight from Winnipeg to Manchester, during a stay in England of three weeks, and then after return to Winnipeg; a 6-hour time shift was involved. The results of this study were analysed by fitting sine curves to the excretory data for the different constituents and thus determining objectively the time of peak excretion together with the standard deviation of the estimates. The results for potassium excretion were essentially similar to Sharp's on subjects in the Arctic. For the first day or so after travel the rhythm was often disorganized and it was impossible to fit a sine curve; there was then an immediate partial adaptation of some 3 or 4 hours while the remaining adaptation of peak excretory time might take up to a week. After the return to Winnipeg the readaptation followed a similar time course (Fig. 9.4).

Sodium excretion showed a similar tendency, but was more irregular than that of potassium so that sine curves could seldom be fitted. The morning fall in phosphate excretion appeared to shift immediately from Winnipeg to Manchester time and back again, but the shape of the excretory curve was not sinusoidal, since the maximum and minimum were not usually separated by 12 hours.

In another study, on a female subject before and after a flight from Manchester to Chicago, which again involved a 6-hour shift, a wider range of variables was measured.

Potassium excretory rhythm on the day after the flight was still adhering fairly closely to Manchester time; on the next day the maximum occurred rather later, and by the fourth day it was adjusted to Chicago time. Its amplitude remained normal throughout. The rhythms of sleep took a similar time to adjust; for the first four days the subject felt very tired around 18.00 hours, which would correspond to midnight in Manchester, and had broken sleep at best after 02.00 when by Manchester time it was 08.00 hours, but by the fifth night she was sleeping fairly well. A simple perceptual-motor test took about 4 days to adapt, showing on the third day two peaks of performance corresponding to those

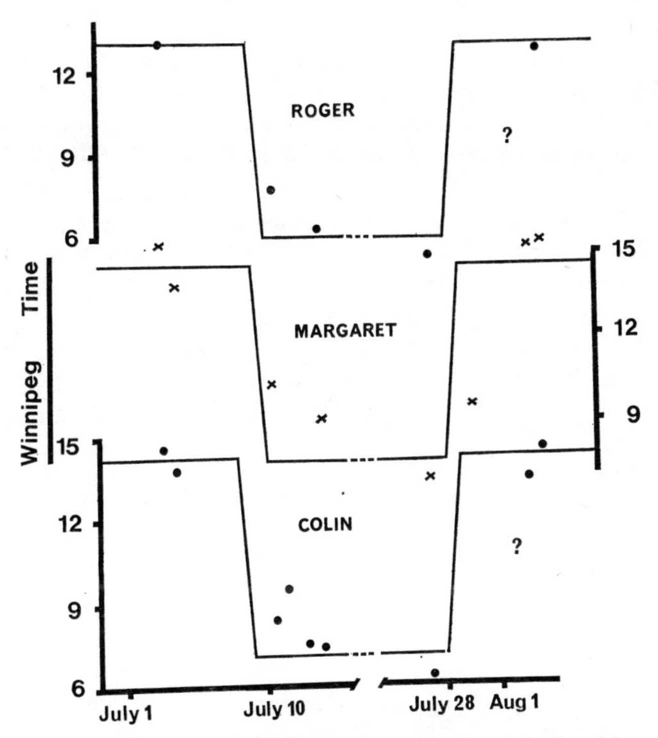

FIG. 9.4. Time of maximum potassium excretion, calculated from best-fitting sine curve, in each of three subjects flying from Canada to England, a 6-hour time shift, and back. Continuous line indicates the change in clock time.

to be expected on Manchester and on Chicago time. Temperature rhythm was not fully synchronized with Chicago time on the fifth day but was fully entrained on the eleventh. The adrenals, as has so often been reported in other situations, were the slowest to adapt. The normal pattern was completely distorted for the first eleven days, in that the plasma concentration of 11-hydroxycorticosteroids remained high at all of the 6-hourly sampling times, failing to show the customary fall to low values in the late evening and early morning; lower values and an appropriate rhythm were first seen a week later, 18 days after arrival. The dissociation between the entrainment of the steroid and the renal potassium rhythms is further evidence against the adrenals playing a critical role in determining the urinary excretion of potassium.

In summary, studies upon abnormal time schedules have provided massive support for the idea that many physiological rhythms are endogenous, but have thrown only a fitful and wavering light upon their possible causal interconnexions; adaptation to a phase shift stands out as much easier than adaptation to a different cycle length; although an endogenous rhythm with a new phase can be developed, with the clock itself seemingly adjusted, there is serious doubt whether the same can ever be achieved for an altered cycle length; under favourable conditions one may expect most rhythms to adapt to even a large phase shift within a week, but the adrenals are commonly much slower. A claim (6) that the influence of the previous time schedule could still be discerned in the potassium excretory rhythm 63 days after a flight from India to England goes, however, far beyond the claims of any other workers in this field. What importance these rhythms may have for those who are obliged to alter their time schedules will be considered in the next chapter; but it is well established that under identical conditions some subjects can adjust their rhythms more rapidly than can others.

Chapter 10

Applied Aspects

Shift-working—Reaction of Workers to Different Shift Systems—Health of Shift-workers—Persistence of Circadian Rhythms in Shift-workers—Merits of Different Shift Systems—Time-zone Transitions—Astronautics.

This chapter deals with two very practical aspects of human circadian rhythms—shift-working and the effects of travelling across time zones.

SHIFT-WORKING

People who work on shifts or at night and thus have to live on abnormal time routines have been the source of much information about human circadian rhythms, as we have seen in previous chapters, but they also pose increasingly important problems in applied physiology. In Britain alone the proportion of shift-workers in industry is estimated to have risen from 12 per cent of the total labour force in 1954 to 18 per cent in 1964 (Ministry of Labour, U.K.), (624).

Many studies on shift-workers have been concerned with sickness and other absence from work, satisfaction with shift-working, or preference for one or another arrangement of shifts, or working efficiency. In only a limited number of these studies have physiological rhythms been explicitly observed. Many factors contribute to any recorded satisfaction or dislike: total hours worked, wages earned, length of annual holidays, leisure time with wife and family or when shops, transport and amusements are uncrowded. It is usually impossible to show conclusively how far any conflict between circadian rhythms and working hours contributes to dissatisfaction or maladjustment. Thus poor sleep at unusual hours may be due to failure of synchronization with endogenous rhythms or to noise or light, or the knowledge that children are at home and would welcome their father's company.

REACTIONS OF WORKERS TO DIFFERENT SHIFT SYSTEMS

The practical implications of physiological rhythms for night-work first became the subject of intensive investigation during the 1914-18 war. In Britain a Health of Munition Workers Committee (625) found that with fortnightly changes of shift, absenteeism was less and output higher during the first week of night shift than in the second week. In a final report (626) the same committee considered that there was no significant difference in output between men on alternate day and night shifts as compared with those on continuous night work. A further report, by Vernon, appeared at the beginning of the second world war (888). This war also, of course, stimulated work on immediate military aspects and some interesting studies on naval personnel have been reported (467, 474, 880). A customary routine on surface vessels was found to be a watch of 4 hours followed by 8 hours off, with morning and evening 'general quarters' and 'dogging' the watch so that the 16.00-20.00 hours watch is divided into two groups with the second going on at 18.00 hours. Physiologically this appeared to be a very disadvantageous arrangement: alertness, temperature and sleep/wakefulness cycles were found to follow the normal pattern, and at night when alertness was most required most of the crew were asleep and those awake were at their least efficient.

Submarine crews not required to attend 'general quarters' (an hour's morning and evening standing at battle stations), not subject to reveille, permitted to sleep in daytime and free from awareness of the alternation of day and night, had a much more favourable routine. Even so the 8 hours off system meant that a maximum unbroken sleep could not be of more than 7 hours duration and alertness throughout the 24 hours was uneven. During the changes of watches, crew were often required to be on duty within thirty minutes of being awakened. Out of 74 members of a submarine crew questioned by Kleitman (467) nine claimed to be fully alert immediately on awakening, another 21 in less than fifteen minutes, 44 required twenty minutes or more and of these 14 needed 1 to 2 hours. Even these figures were viewed with some doubt by Kleitman who considered that the crew tended, if anything, to overestimate their ability to become fully alert at short notice.

A study by Wyatt and Marriott (941) on night work and shift changes in three factories having a weekly, a fortnightly and a monthly change of shifts respectively, is of particular interest in this respect. Output was found to be slightly lower on night work than on day work with all three types of shift change. Some individuals, however, had as high, or even a higher, output on night work. Output increased in the second week on day shift when shifts were changed fortnightly, and in successive weeks on day shift when shifts were rotated monthly. Absence from work on day and night shift when the same men were compared showed an increased absence with successive weeks of night shift and a decreased absence with successive weeks of day shift. Most of the men expressed a preference for day work and feelings of fatigue were stated to be more prevalent on night work. Forty-two per cent of the men stated that they were unable to get sufficient sleep when on night shift and 62 per cent took one or more days to adapt to changes in meal times, 23 per cent not settling down to a new shift for more than six days. The attitude of wives to their husbands working continuous shifts is also very important in considering any system of shift work as Banks has emphasized (39). Working over the weekend has been found to be a particularly frequent and persistent objection to shift working (137).

A permanent night shift, manned by volunteers, has been reported (183). A bimodal distribution of the workers' preference suggested a preference for either permanent day or night duty. The night duty worked was, however, alternate nights. Ten out of fifteen night workers questioned mentioned some aspect of regularity as an important reason for their preference for night duty. In a plant where, after a ballot among the workers, the system was changed from a continuous (7 shifts) system to a $3 \times 2 \times 2$ cycle, it was found on interviewing 50 workers 3 years afterwards that 43 preferred the shorter cycle (de la Mare and Walker), (184). There were no advantages in terms of pay or total number of hours worked. Only four of those who preferred the longer cycle were stated to have given reasons indicating adaptation during the continuous shift.

The effects of a changeover in a steelworks from a discontinuous three-shift system with breaks at weekends to a continuous $2 \times 2 \times 2$ system have been described by Walker (904) and also contrasted with the $3 \times 2 \times 2$ system in a chemical works (see also de la Mare and Walker, (184), above). In the $2 \times 2 \times 2$ system rotation always took place after two shifts and there was an 8 week cycle. The change had been made at the men's request 1 to 3 years previous to the investigation. It was accompanied by a decrease in working hours from 45 to 42 hours weekly and a 20 per cent increase in earnings and was, therefore, weighted in the direction of a favourable response to the change. Sixty men were interviewed. All preferred the new cycles and more than half the men considered that sleeping arrangements were preferable with the new system. They claimed that they had more time to rest, were not so tired and that a 24 hour break on changing shift enabled them to recover from fatigue better and that they felt fresher. Major advantages of the new system were said to be the opportunity to enjoy more frequent, even if shorter, spells at home, at normal times every week and being able to get out every week. This was contrasted with the monotony of seven consecutive shifts and the 'dead fortnight' of the consecutive afternoon and night shifts. Loss of the weekend breaks was, however, a disadvantage. The chief objection to the system was, in fact, the curtailment of the long break after a night shift by the need to sleep on the first day. Putting the long break after another shift would, however, bring the more arduous shifts within 24 hours of one another. Both wives and men found little difficulty in understanding and following the $2\times2\times2$ shift whereas the $3\times2\times2$ cycle was still causing misunderstanding three years after its introduction in the chemical works. There was no evidence of either an increase or decrease in absence from work due to the change in the system of shifts. The change in shifts described by Walker was initiated by the workers. The management was agreeable to the changes but raised the possibility of certain disadvantages. These included the question of difficulties in administration and communication, and of increased absenteeism. These difficulties did not, however, arise in practice.

HEALTH OF SHIFT-WORKERS

The occurrence of illness such as peptic ulceration, sleeplessness and nervous disorders in shift- or night-workers has been the subject of a number of recent papers (1, 2, 73, 127, 197, 207, 647, 660, 859, 866, 900, 941). A greater occurrence of gastrointestinal and neurotic illness has been reported (1, 2), though a careful statistical study (207) gives no indication of an increased incidence of peptic ulceration in shift-workers.

Sickness absence is stated to be lower in shift-workers than in day-workers (900). Aanonsen (2) studied the rates for sickness absence amongst workers in a chemical plant. 339 subjects were on day work, 372 on shift-work. The shift-workers had less illness than the day-workers but detailed statistical analysis was not performed.

Taylor (859) in a very thorough and carefully evaluated study of sickness absence amongst workers in an oil refinery, found that shift-workers had consistently and significantly lower rates of sickness than comparable day-workers. A 42 hour 5-day week was worked at the plant, which was in continuous operation. Shift-work was arranged on a weekly three cycle system, 06.00-14.00 hours, 14.00-22.00 hours and 22.00-06.00 hours. Three days were allowed off after each period on morning or night shift with an additional extra day off during the course of every four weeks so that an average 40 hour week was worked. Morning shifts began on Tuesdays, afternoon and night shifts on Fridays. Shift-workers were paid a basic 30 per cent extra per week and, by suitable adjustment of their shift, could arrange three weeks annual holidays instead of two. Day-workers usually worked two to three hours overtime per week. Shift-workers were obliged to wait a minimum of four extra hours if their relief had not arrived and sometimes worked 12-hour shifts when one worker in a group was absent. About half the hourly-paid workers had been on shift work for some years. Shift-workers usually were part of a small, two to six man, group. Day-workers were normally in groups of well over ten men. Sick pay was at full normal wages for a period of from 3 weeks to 27 weeks according to length of service plus a further and similarly adjusted period at half pay.

Crude sickness rates were measured for all day and shift-workers throughout each of the years 1962, 1963, 1964 and 1965 (between 1,301 and 1,718 workers per year). The sickness rate for shift-workers was less than that for day-workers as assessed by a number of different criteria.

Interviews with 150 shift-workers were carried out and over 75 per cent expressed a preference for shift work. The reasons given were usually similar—regular schedule, less need for over-time, opportunity to enjoy hobbies or go out in daytime at offpeak periods and to travel to work when traffic was not congested. Those who stated a preference for day work generally mentioned time off at weekends or in the evenings or had wives working fulltime on day hours.

Taylor suggests that the reasons for the reduced rates of sickness and other absence amongst the shift-workers whom he studied may lie in the small size of the shift-workers' groups, the preference for the hours worked and a higher level of satisfaction with their job and identification with it.

PERSISTENCE OF CIRCADIAN RHYTHMS IN SHIFT-WORKERS

It is obvious from the foregoing that many factors apart from physiological rhythms contribute largely to satisfaction or discontent with shift work. There are, however, a number of investigations of physiological circadian rhythms in shift-workers, and some indication that these may be of importance. Perhaps the most obvious of these is the rhythm in measurable performance, which at various tasks has been found to vary quite markedly during the 24 hours. Browne (121, 122) has shown how teleprinter switchboard operators working at a large communications centre had a clearcut circadian variation in performance. The time taken to answer a call and the total number of calls being made were measured continuously over a period of three months. Performance as judged by delay in answering calls was found to improve from 09.00 hours to midday, remain fairly constant until 17.00 hours and then decline increasingly quickly to 04.00 hours. There was then a slight improvement followed by a further fall until 08.00 hours. During the day pressure of work did not affect performance within each hour but at night there was a slight but consistent tendency for the delay to fall when the pressure of work increased. Even at night, however, load had much less effect on performance than had time of day. The evening (16.00-23.00 hours) shift was found to be most efficient, the night (23.00-08.00 hours) the least efficient.

Errors in meter reading and computation have been studied very intensively by Bjerner, Holm and Swensson (72). The subjects concerned were reading instruments which measured such information as temperature, gas consumption and gas pressure. Entries of the resultant readings and calculations were made hourly. At 06.00 hours the entries were checked and any errors corrected. Over 175,000 ledger entries were examined. A distinct circadian variation was found with two peaks in the error reading, a large peak at 03.00 hours and a lesser one at 15.00 hours (Fig. 10.1). Changes of shift, which were at 06.00, 14.00 and 22.00 hours, had little influence. The variation in number of errors was also found to be similar during the first and last three nights of a night shift; and a reduction in the length of the working week, though reducing the number of errors, left the temporal pattern unaltered.

Industrial accidents have also been found to show a 24-hour variation. In a study of railway employees Menzel (582) noted a peak accident frequency between 22.00 and 02.00 hours.

Other examples of variations in reaction time, alertness and wakefulness throughout the 24 hours are discussed in Chapter 7

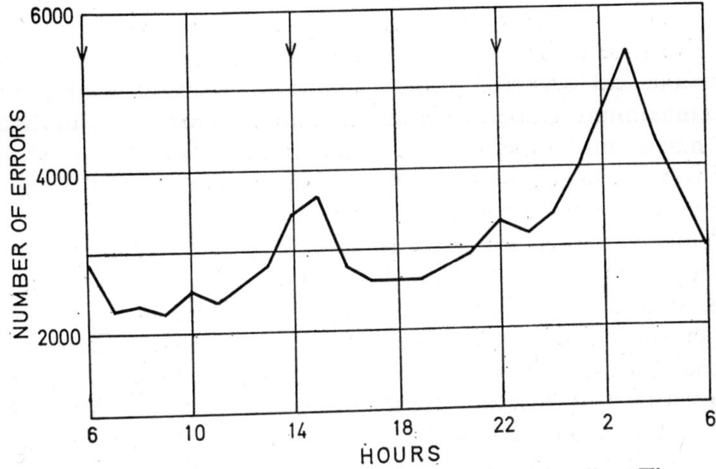

Fig. 10.1. Circadian variation in errors in meter reading. Time hours is plotted along the abscissa and errors along the ordinate. (Bjerner, B., Holm, A. and Swensson, A., Fig. 1, 72.)

and their relationship to underlying physiological processes is emphasized. As was described in this chapter, temperature, in a great variety of circumstances, correlates well with performance and is much easier to measure. Jaeger (422) as long ago as 1881 noted a reversal of the temperature rhythm in bakers who worked through the night. Benedict, 1904 (58), did not find clear inversion of the temperature curve in a nightwatchman whom he studied but did observe that the normal pattern was lost. Toulouse and Pieron (874), however, investigated the position in six subjects who had been nightwatchmen for several years and found inversion of the temperature curve with a fall in temperature during the afternoon and evening.

Temperature rhythms in nurses who had a month's spell on night duty every three to four months during which they worked from 20.00 to 08.00 or 09.00 hours have been compared (860) with those of workers in a bakery. The nurses in the first week of night work showed only a slight trend towards inversion of the temperature curve and even at the end of the fourth week of night duty no nurse had a completely inverted temperature rhythm. Two of the three bakery workers who were stated to have always done night work were found to have a complete inversion of the temperature rhythm, and the third had his maximum temperature at 02.00 hours. Another 13 bakery workers were well accustomed to night work but sometimes worked on day shifts for a fortnight on end. In 9 of these 13 temperature inversion had occurred within 5 days of going on night shift with maximum temperatures at 07.00 hours. Three more of them had maximum temperatures at 02.00 and one only did not show an inversion of temperature. Reversion on returning to the day shift was sometimes even more rapid than inversion.

Temperature records obtained at 4-hourly intervals from three subjects during a week of working on day shift, thirteen weeks of night shift, and another week of day shift have been recorded by Bonjer (87). Their temperature rhythms reverted to a normal diurnal pattern during each weekend off work, and progressively adapted to a nocturnal pattern during each week on night shift. Two of the subjects adapted in this way within the first two days of the week, the other subject took four days at first but after four weeks of night shift became adapted on the third day. Similar findings have been reported (Fig. 10.2) by Van Loon (886) who noted also some indication of long-term adaptation to night work in that, over several weeks, the return to nocturnal timing after the weekend became more rapid.

Plasma corticosteroid concentration has also been supposed (321) to be associated with alertness, and persistence of the normal corticosteroid rhythm in nurses and nightwatchmen has been reported (598). We have found, however, that adaptation may occur in this rhythm in subjects on habitual night work in an industry where night working predominated and the social background is adjusted accordingly, by contrast with night shift workers in a basically day-orientated industry (158).

MERITS OF DIFFERENT SHIFT SYSTEMS

If one tries to recommend the best arrangement for shift-working, the physiological and social considerations conflict. Teleky (860), on the basis of the time taken for the temperature curve to adapt to night work, recommended for factory workers a rotation at longer than weekly intervals and preferably at monthly intervals, and for other workers, such as nurses, rotation at even longer intervals.

The usual custom at present in many industries: to work about a week on one shift, have a rest break, and change to another shift, is at complete variance with this recommendation. It is perhaps

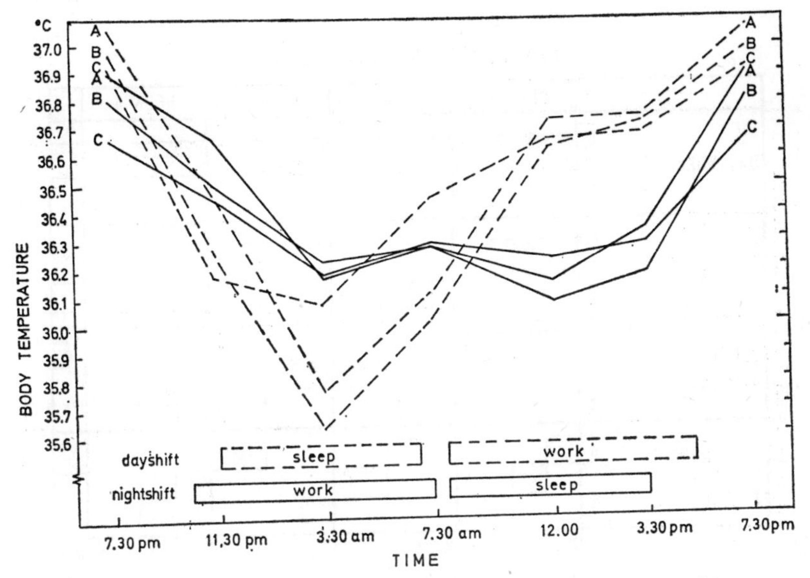

FIG. 10.2. Body temperature curves of three subjects on day shift (interrupted lines) and on night shift (continuous lines). The readings over a period of two or more weeks have been averaged for each subject. (Van Loon, J. H., Fig. 1, 886.)

the most unsatisfactory arrangement that could be devised, for just as the subject's rhythm has been thoroughly disturbed and he is beginning to entrain to the new routine, he is switched to a different schedule (647). Either extreme is better, though for different reasons. Continuous working on night or any other shift permits the maximum of adaptation of all physiological rhythms, particularly in an isolated community such as the Spitsbergen coalminers (534) where the social life is geared to the working routine. For most people living in a community who follow regular nychthemeral habits this is, however, unacceptable on social grounds.

The opposite extreme, the very frequent change of shift (Fig. 10.3), is being increasingly introduced and seems to be the most acceptable to workers. It permits a wider range of social activity, at fairly frequent intervals; and though there is no possibility of adaptation of physiological rhythms to suit the hours of work, the basic nychthemeral rhythm probably persists and leaves the worker properly synchronized for his leisure habits. It would be interesting to know whether, after long periods on such rapidly

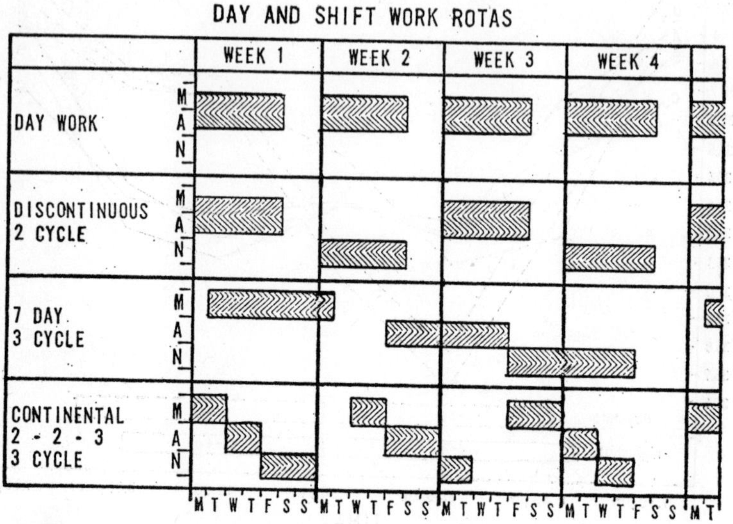

Fig. 10.3. Various forms of shift work rotas. (From data of Dr. P. J. Taylor.)

alternating shifts, the normal endogenous rhythms weaken or disappear, as the renal component has been reported to do in men working entirely irregular hours (538).

TIME-ZONE TRANSITIONS

Journeys across time-zones are also interesting both in providing information towards understanding circadian rhythms and in the practical problems which they pose.

Here again it is important to distinguish between the specific effects of altered time, and other consequences of travel and arrival in a society differing in both obvious and subtle ways from that which the traveller left. The only deliberate attempt to dissociate these factors comes in a comparison between effects of eastward, westward and southward flights (362, 363, 364). Adaptation on westward flights is stated to take longer than on eastward flights (629).

In 1939 the traveller by the fastest liner took several days to reach New York from Europe and his physiological rhythms adapted without obvious difficulty. Nowadays an air journey from London to New York takes only about 8 hours and with the advent of supersonic transport this will be shortened still further; the Concorde, for example, is expected to fly the Atlantic in less than 3½ hours. The long-distance air traveller finds himself quite suddenly in a time zone completely out of alignment with his biological processes (Fig. 10.4). Sleep and wakefulness, digestive processes, body temperature, and mental activity may all show quite marked disturbances which, even subjectively, may be noticeable for some days after arrival. The problem is important for aircrew, in whom it may be further complicated by successive transitions backwards and forwards, though sometimes it is minimised by an immediate return to the original time zone. It is also important to executives, military personnel and other individuals concerned in decision-making whilst still suffering from the disorientating effects of the time-zone transition. It has been found (477) that variations in reaction time appear to be closely related to the body temperature (see also Chapter 7). Circadian rhythms in mental alertness may also be considerably disturbed as one of our colleagues has shown (259). From our personal enquiries it is clear that the effects of time-zone transitions are now a matter of serious concern in many international companies, military forces and diplomatic and political circles.

Strughold (847) observed his own sleep and eating habits following a flight from the U.S.A. to Germany and after the return journey. He considered that it needed at least a week for these to adapt to the new time zones. Other studies on the temperature rhythm indicate a similar period of disturbance after long-distance flights (133, 161, 164, 789). We have found that even more prolonged disturbances in plasma corticosteroid levels may occur and, by implication, in the activity of the pituitary function and higher centres also (164).

An extensive study, but subjective only, on the effects of time-zone changes on sleep and digestive functions has been made by

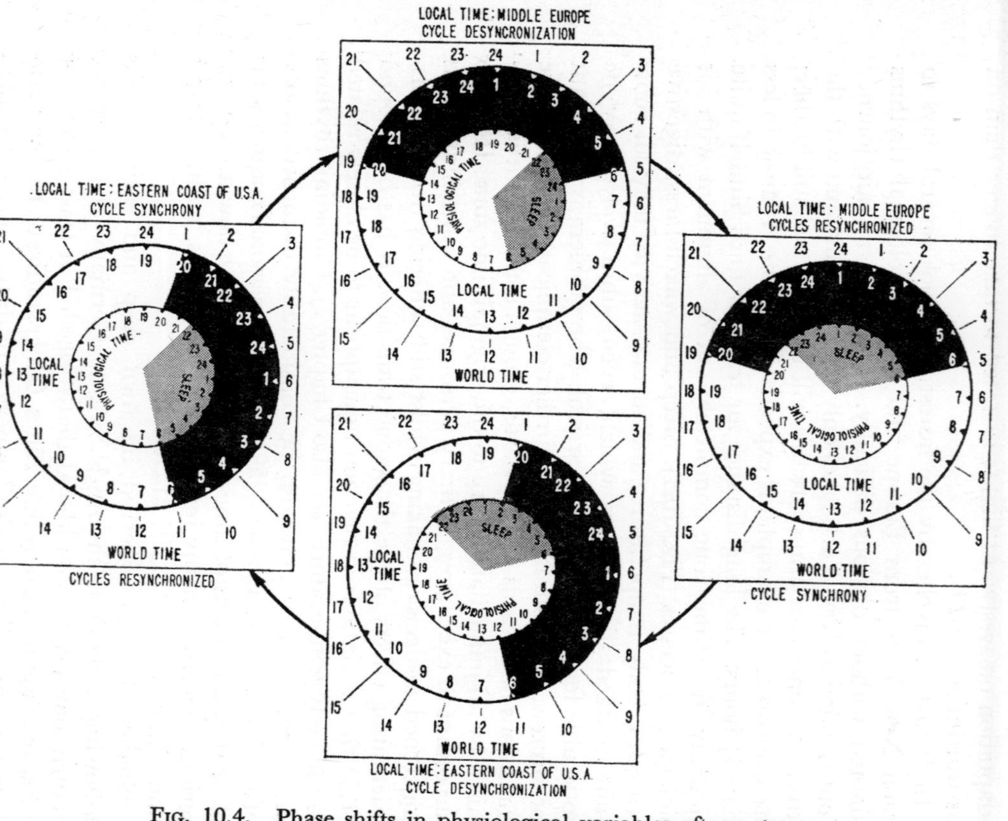

FIG. 10.4. Phase shifts in physiological variables after a transatlantic flight. (Strughold, H., Fig. 2, 849.)

Lavernhe, Lafontaine and Laplane (502). Questionnaires were sent to 847 flight personnel of an airline's 'long distance' flight department. 312 subjects replied, the highest percentage of answers being provided by navigators and captains. The majority (76 per cent) could not sleep normally on the third night or even later. 41 per cent complained of digestive disorders. Irregular time displacements such as a flight from Paris to Tokyo and back followed two days later by a Paris-Los Angeles flight, were particularly disturbing sometimes causing, it was stated, fatigue to a degree that could only be reversed by prolonged rest.

A relationship between time of flight departure and strain and performance in flying has been suggested by Klein *et al.* from studies of pilots making transatlantic flights (463).

Few scientific attempts to mitigate or obviate the effects of time-zone transitions have been reported. Gerritzen (281) studied the effects of inverse illuminations on the renal excretion of five students remaining at strict bed rest. Their sodium, potassium and chloride excretory rhythms were reversed in phase but very flat, and no more informative components of circadian rhythmicity were recorded.

The information available on the effects of rapid time-zone transitions on bodily functions is still relatively scanty but is sufficient to indicate that the physiological and psychological effects of such journeys are considerable and may be rather more long-lasting than is at present fully realised. There would, however, appear to be no evidence whatever that any serious illness is directly associated, though obviously disturbances of sleep and fatigue, and stress phenomena, may be exacerbated by such flights.

The extent to which the performance and mental ability of aircrew or others travelling frequently is disturbed by these transitions is, however, a more open question, and does not appear to have been the subject of much detailed scientific investigation. Schreuder (803) in a review of medical aspects of aircraft pilot fatigue, suggests that some degree of adaptation to time zone transitions may be shown by airline pilots. This is highly likely but its existence or extent is at present only a matter of conjecture as is its sufficiency to compensate for the normal variation in wakefulness and alertness during the twenty-four hours. The already very stringent regulations on flight time of British pilots now include the desirable provisions that where a time zone change of more than four hours is involved, or the flying duty period extends through 03.00 hours local time at destination, the subsequent rest period should not be less than twelve hours (84).

Advice on steps to avoid or mitigate the effects of time-zone transitions has been difficult both because of our limited knowledge of the underlying physiological processes and because the most effective advice, which is to take a few days rest after arrival, is clearly impracticable for most people. Strughold (849) has suggested that some such adjustment, which he termed post flight pre-adaptation, may be considered if full alertness is required on a special occasion or, alternatively, that the subject may, before his journey, adopt a sleep/wakefulness pattern corresponding to that of his destination (pre-flight synchronization). Administration of drugs he suggests as a possible third method, and the discreet use of sedatives and stimulants at appropriate times would certainly seem worth considering. McGirr (661) advises in more practicable if less effective terms that when a journey across five or more time-zones is being made a flight should be chosen whose arrival coincides with the usual bedtime, and that on reaching his destination the subject should go to bed, taking a sedative if necessary. He also recommends the avoidance of business meetings or entertainment on arrival. From our personal enquiries it would appear that in some companies the effects of time zone transitions are taken into consideration and executives are advised not to enter into serious negotiations until they have had 24 hours to adapt to the new environment and equally not to attempt to return to the office immediately on arriving home. In many companies, however, personnel are either required or expected to ignore the effects of such journeys and to conduct negotiations abroad whilst at a distinct disadvantage, and equally to write reports immediately on return irrespective of how inferior these may be when made under such circumstances. The saving in time by such procedures may well be very expensive.

Statesmen face similar problems and since their decisions may well be so much more important, it would seem advisable that the effects of rapid time-zone transitions be emphasized to them by their advisers, and that where possible fatigue and disorientation should be lessened by such measures as ensuring a night's sleep on arrival.

CIRCADIAN RHYTHMS IN ASTRONAUTICS

Circadian rhythms, particularly in sleep and wakefulness, are very important in space flight. The Russian cosmonauts have shown a sleep/wakefulness cycle in which their periods of sleep have coincided with night-time in Russia (848, 849). Bykovsky is reported as having four 8-hour periods of sleep (692). Tereshkova showed a regular circadian rhythm in physiological functions except during the first and last days. The crew of three men in the Spaceship Voskhod had a shift system of sleep and rest (277). Pre-flight isolation tests in preparation for such journeys are known to have been carried out by the Russians (828).

A number of experiments on work-rest schedules in a space vehicle simulator have also been performed in the United States (7, 12, 13, 310). During periods of 15 days confinement, schedules of 4 hours work/2 hours rest were examined, and on 30 days confinement a schedule of 4 hours work/4 hours rest was followed. Performance, as measured by arithmetical computation, was better on the 4 hour work/4 hour rest schedule than on the 4/2 routine. Heart rate and body temperature remained on a circadian cycle on both schedules. Physiologically neither routine would appear advisable.

The crews of the GT4 and GT5 space missions are stated (849) to have noted no difficulty in performance attributable to the 90 minute orbital cycle of light and darkness to which they were exposed. McDivitt and White, however, in the GT4 spacecraft had sleep periods which were not always related to night time at Cape Kennedy, their point of origin, and lost sleep and became fatigued during the four-day mission. Cooper and Conrad in the GT5 vehicle were each allowed a long sleep period corresponding to night time in Cape Kennedy and tended to stay on their pre-existing sleep/wakefulness schedule. Great care in the scheduling of sleep/wakefulness cycles would seem to be a fundamental requirement of such missions.

Subjects escaping from the gravitational field of the earth, as in moon flights or the projected interplanetary travel, will be exposed simultaneously to day and night, to a black sky and to constant sunshine. On the moon they encounter the twenty-seven day selenographic cycle with illuminance from the earth seventy-five times that of the full moon (849). During the Apollo moon missions the crew's circadian rhythms were disrupted and the astronauts became fatigued to a degree which was described as producing 'a most unsatisfactory situation in flight' (65).

The practical implications of human circadian rhythms are as yet largely in the exploratory stage. With ever increasing emphasis on round-the-clock utilization of capital equipment together with a shortening working week and consequently increased shift working, and a continued expansion in leisure and business travel, they are likely to become ever more important.

Chapter 11

Clinical Implications

Temperature—Endocrinology—Respiratory, Cardiovascular and Renal Disorders — Glaucoma — Epilepsy — Psychiatry — Obstetrics — Malignancy — Parasitic Infestation — Pharmacology and Therapeutics — Mortality and Susceptibility.

The clinical implications of circadian rhythms are gradually coming to be realized as our knowledge of these processes increases. For example, a biochemical estimation or other physiological measurement at one time of day may differ markedly from one made at some other time during the course of the 24 hours. Many examples are mentioned in previous chapters. Similarly a drug given at a particular time may have effects of a very different order of magnitude from those when it is administered at some other hour of the day or night. Many other aspects of physiological rhythms may be disturbed in disease, and a comprehensive account of all those which have been claimed would be beyond the scope of this book. In this chapter we describe a few examples of circadian rhythmicity in various aspects of medicine, and the references will permit a more extensive study.

TEMPERATURE

The interest of clinicians in body temperature as a diagnostic aid has produced numerous charts showing the circadian rhythm in body temperature in afebrile patients. Such modifications as a rise in the mean level with persistence of the normal rhythm, or an increased amplitude of the rhythm, have been noted in different forms of febrile illness. Clinicians have thus for long been accustomed to consider the pattern of body temperature rather than relying upon a single reading. Similarly, when temperature is used as a guide to the time of ovulation it is essential to pay attention to the circadian variation, as for example by considering only readings taken first thing in the morning (691). More complex disturbances of body temperature have been recorded in epileptics and in sufferers from malignant neoplasms, but the association is not sufficiently clear cut to have, as yet, any great diagnostic or prognostic value (322, 760).

ENDOCRINOLOGY

ADRENAL CORTICOSTEROIDS

Clinicians are well aware that plasma corticosteroid concentration varies circadianly in healthy people. Cushing's syndrome provides a good example of disturbance of this rhythm. Laidlaw *et al.* (495) appear to have been the first to notice that in this condition the normal corticosteroid circadian variation is not present. High day and night plasma 17-OHCS levels in three patients with Cushing's syndrome were reported soon afterwards (526) and since then absence of the normal circadian rhythm in plasma corticosteroid levels in Cushing's syndrome has been noted by many observers (111, 202, 204, 226, 531, 554, 672, 678). Doe *et al.*, 1960 (205), in a study of five patients with Cushing's syndrome due to bilateral adrenal hyperplasia, found that the subjects had constant levels of plasma and urinary 17-OHCS throughout the 24 hours and that Na and K excretion in these patients was also disturbed with a nocturnal rise. The suggestion has been made that in some cases of Cushing's syndrome in which there is a loss of the normal circadian rhythm and persistently raised levels of plasma 17-OHCS, a relatively constant secretion of a small amount of ACTH may be the underlying cause.

Ekman *et al.*, 1961 (226), studied a group of twenty-two patients with Cushing's syndrome, one of them a woman who, after a partial adrenalectomy five years previously, had an apparently total remission. A series of blood samples was obtained over the 24 hours. Urinary 17-ketosteroids and 17-ketogenic steroids were also measured. The majority of patients showed high nocturnal plasma 17-OHCS levels together with an absence of the normal circadian variation. In only four patients were plasma corticosteroid levels below 10 μg/100 ml noted at any sampling time, and in only three was there a marked fall during the night. Even in the patient who appeared clinically in a state of remission plasma corticosteroid levels were fairly constant. The authors suggest that non-neoplastic Cushing's syndrome may often be due to a disturbance in the hypothalmic control of ACTH. In most subjects urinary excretion of 17-ketogenic steroids was increased and was usually paralleled by a rise in 17-ketosteroid excretion. In a number of subjects, however, many ketogenic steroid estimations were within normal limits; three cases, in particular, showed no increase in 17-ketogenic steroid excretion despite repeated examination. The importance of making serial measurements of plasma corticosteroids in the investigation of suspected cases of Cushing's syndrome, even though urinary excretion of cortisol metabolites is normal, is emphasized by these findings. Estimations at night may be more important in diagnosis than those made from samples obtained during the day (109).

A normal circadian rhythm in plasma 17-OHCS in a small group of three thyrotoxic subjects has been reported (705). A careful study, however, of twenty-one patients, thirteen with hyperthyroidism and eight with myxoedema, suggests that some changes in the steroid circadian rhythms occur in thyroid disorders (573). In normal subjects the concentration of conjugated 17-OHCS in the plasma lags behind the concentration of free steroids by 2-4 hours, and the urinary excretion reflects with a slight delay the plasma concentration of conjugated steroids. In thyrotoxicosis the concentration in the plasma of free steroids is low and of conjugated steroids is high, and the noon peak of plasma concentration of conjugated steroids and of urinary excretion is both high and somewhat delayed. The converse is observed in myxoedema, where the concentration of free steroids in the plasma is high; that of conjugated steroids is low with little discernible rhythm, as is also the urinary excretion. It seems that steroid production is still under the control of ACTH, but that the thyroid accelerates the conjugation and consequent excretion, as has been seen also when the adrenal cortex was stimulated with exogenous ACTH (705). The changes characteristic of hyperthyroidism could not, however, be precisely replicated by giving tri-iodothyronine to normal subjects (573).

Absence of the normal circadian rhythm of adrenocortical function in cases of advanced bronchogenic carcinoma has been reported (817) but other workers have failed to confirm this observation in samples taken at 08.30 and 16.30 hours (520).

As has been considered in Chapter 3, the rhythmic secretion of the adrenal cortex is commonly held to be responsible for many other aspects of physiological rhythmicity. The evidence for adrenal involvement is, in part, the absence of other rhythms in patients with adrenal deficiency, or with no adrenal rhythm as in Cushing's syndrome. The morning secretion of cortisol has also been implicated in the periodic manifestations of other disease, such as the rhythm of haemolysis in paroxysmal nocturnal haemoglobinuria (643, 644).

DIABETES MELLITUS

A number of periodic variations in metabolites in diabetes mellitus have been described (638), including a 24-hour rhythm in citric acid excretion and in β-hydroxybutyric acid excretion and ammonia production in the kidney.

Möllerström (632-637) has extensively studied the circadian rhythms of carbohydrate metabolism in diabetes mellitus and has suggested that in severe cases the dosage and timing of insulin injections should be related to the rhythm of β-hydroxybutyric acid excretion. The need for morning injections of short acting insulin preparations for adequate control of insulin-dependent diabetic patients may be explicable in terms of morning insulin/glucose ratios being higher than afternoon ratios (266). It is known also that significantly higher a.m. than p.m. plasma glucose values occur in diabetics (240). However, although a number of observers have claimed that the fasting plasma triglyceride levels may be an indicator of its degree of circadian variation (397, 574, 845) the fasting blood sugar in insulin-treated patients is not a reliable indicator of the circadian variations in blood sugar (574).

These metabolic rhythms in diabetics, and the uncertainty how far they are dependent upon habits of sleep, work and meals, are a serious contra-indication to the employment of diabetics on shift work; they also suggest the need for caution when diabetics for any reason depart from the regularity of their habits.

Serial electrocardiograms over the 24 hours in diabetics sometimes reveal a periodic incidence of those changes in the T wave and ST segment which are usually interpreted as indicating coronary insufficiency (458).

RESPIRATORY, CARDIOVASCULAR AND RENAL DISORDERS

Abnormal rhythms in vital capacity in asthma and heart disease have been described (581). Changes in the normal rhythm of vital capacity may occur in patients with pulmonary tuberculosis (198, 200), and many patients have individually characteristic patterns. Complete inversion of the rhythm has been reported in some (199) with a return to normal with clinical recovery.

Forced vital capacity was recorded throughout the day (510) in five healthy men with a mean 1-sec. forced expiratory volume (F.E.V.$_{\cdot 1}$) of 3,270 ml and in sixteen patients with chronic bronchitis and emphysema of whom four were mildly affected with obstructive airway disease (mean F.E.V.$_{\cdot 1}$ 2,190 ml) and twelve severely affected (mean F.E.V.$_{\cdot 1}$ 760 ml). F.E.V. and F.V.C. fell during the night to a minimum around 06.00 hour and then rose during the morning. The nocturnal fall was much more marked in the chronic bronchitic patients than in the normal subjects. A similar circadian variation, with a rise of F.E.V.$_{\cdot 1}$ in the morning and a fall in the afternoon and night has been confirmed in large groups of industrial workers (313, 903) and is independent of the presence of respiratory impairment or of exposure to dust. An exception is found in those exposed to cotton dust (664, 903) in whom F.E.V. falls and airway resistance rises in the course of the working day. The fall was larger in those with byssinosis, and was clearly a result of specific exposure to cotton dust since it was seen during the course of the working shift at whatever time of day or night this was worked, was diminished by installation of dust suppressing machinery, and ceased when workers were no longer exposed to cotton dust. Coalminers exposed to much higher dust levels showed no similar changes.

Haemoptysis is said to be more common during the night (200), as are cerebral haemorrhage, pulmonary oedema, and cardiac asthma (488, 489).

Circadian variations in blood pressure of hypertensives have been recorded (576, 584, 590) and circadian variations in blood pressure may play some part in the occurrence of nocturnal cerebral infarction (195, 873). Patients with acute glomerulonephritis have a marked 24-hour variation in blood pressure which decreases on recovery so that clinical progress may be more clearly noted by taking the blood pressure at the time of day that it reaches its peak (437). Patients with hypertension have significantly greater changes in plasma volume than normotensive subjects and it has been suggested that these variations in plasma volume may be responsible for circadian changes in arterial pressure in patients on sympatholytic drugs (169).

A loss of the normal circadian rhythm of plasma 17-OHCS has been noted in cases of congestive cardiac failure (480). High late evening plasma corticosteroid levels with change of the normal circadian variation have been noted in renovascular hypertension (134). A nocturnal increase in sodium and potassium excretion in patients with congestive heart failure and cirrhosis with ascites has been described (83, 295). Reversion to normal rhythm with normal sodium excretion has been stated to occur in congestive heart failure if a similar posture is maintained throughout the 24 hours (90).

A much reduced circadian variation in haemoglobin levels in patients with advanced heart failure has been observed (732), and could also be due to changes in plasma volume.

In renal disorders the normal circadian rhythms may be disturbed (583, 585) and a high nocturnal output of urine has long been accepted as an indication of disease (841). Renal transplants may be associated with abnormal urinary rhythms (64).

Reversal of the normal rhythm of urine flow and solute excretion in a case of unilateral nephroptosis with a return to the normal pattern on surgical correction of the abnormality, has been reported (209).

GLAUCOMA

Circadian variations in intraocular pressure occur in glaucomatous as well as in normal eyes with higher tensions during the day than at night (96, 97, 220, 453, 501, 671).

EPILEPSY

Langdon-Down and Brain in 1929 (500) noted a rhythmicity in the time of day at which major epileptic attacks occurred in sixty-six patients whom they studied. They considered that the patients could be divided into three groups: those in whom the attacks were mainly diurnal, those who suffered mainly nocturnal seizures, and a third group equally prone either by day or night. The patients in the nocturnal and diurnal groups in the incidence of seizures responded in an opposite manner to the onset and cessation of sleep. These observations have been confirmed and extended by a number of observers (189, 311, 341, 403, 699 and see Fig. 11.1A and Fig. 11.1B).

A steady rise in incidence of seizures was found from 03.00 with a very marked peak between 06.00 and 07.00 hours in a study of 110 patients which covered 39,929 fits which were either major or minor involving loss of consciousness (311). An abrupt fall followed at 08.00 hours. The lowest incidence of seizures was

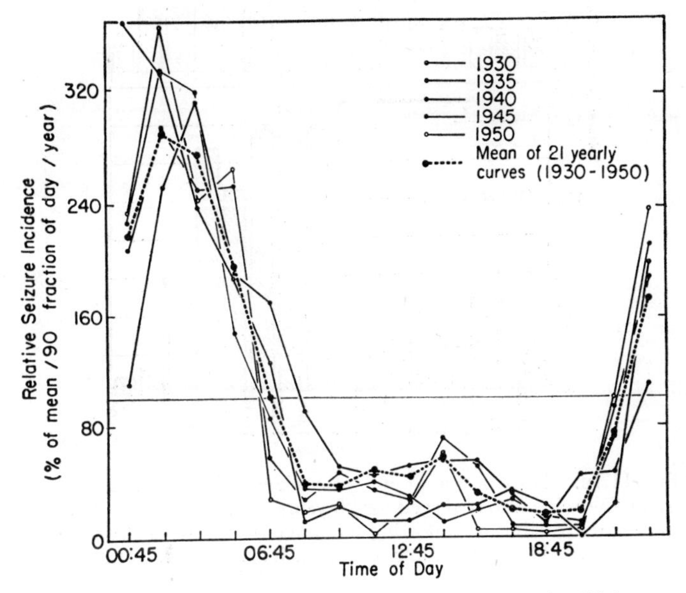

FIG. 11.1A. Distribution of epileptic seizures over the 24 hours in a case observed for 20 years. 'Night type.' (Halberg, F. and Howard, R. B., Fig. 1, 341.)

between 17.00 and 21.00 hours. Smaller peaks were seen during the course of the day and the second highest incidence was between 22.00 and 24.00 hours. A very interesting case was that of a subject who was moved from one institution to another where he went to bed an hour later and began to have his epileptic attacks also an hour later. A tendency for seizures to occur at time of circadian minima in eosinophil levels has been noted and the suggestion made that the incidence of seizures may be related to corticosteroid levels (235, 333). Factory workers on night shift tend to start having their 'nocturnal' fits at the appropriate diurnal sleeptime (60).

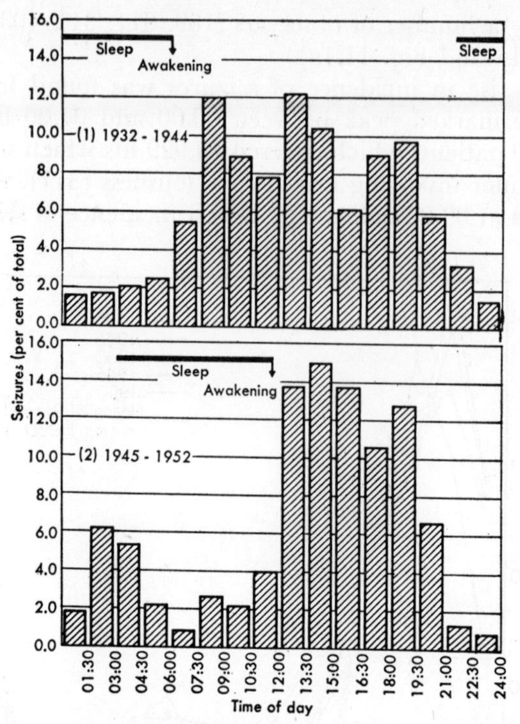

FIG. 11.1B. Distribution of epileptic seizures during the 24 hours in another case also observed for 20 years. 'Day type.' The change in seizure distribution following a change of routine is clearly evident. (Halberg, F. and Howard, R. B., Fig. 2, 341.)

PSYCHIATRY

Changes in circadian rhythms are not infrequently associated with psychiatric disorders. An outstanding example is manic depressive psychosis with daily variations in mood and symptoms and disruption of the normal sleep/wakefulness pattern. Alterations in renal excretory rhythms may also occur in this condition (227, 541). They comprise changes in the amplitudes of the rhythm of water, potassium, sodium and chloride excretion.

Reiss (754) suggested that disturbances of adrenocortical secretion in depressive illness may occur and raised plasma corticosteroid levels in depressed patients have since been noted by a number of observers in morning or evening blood samples (82, 83, 110, 283, 284, 410). Samples obtained at 06.45, 11.45 and

22.00 hours, in an effort to allow for circadian variations, have shown elevated plasma cortisol levels at the three sampling times (206, 658), but other observers who also took samples at three times during the 24 hours have reported exaggerated rhythms in depressed patients as compared with controls and with the same

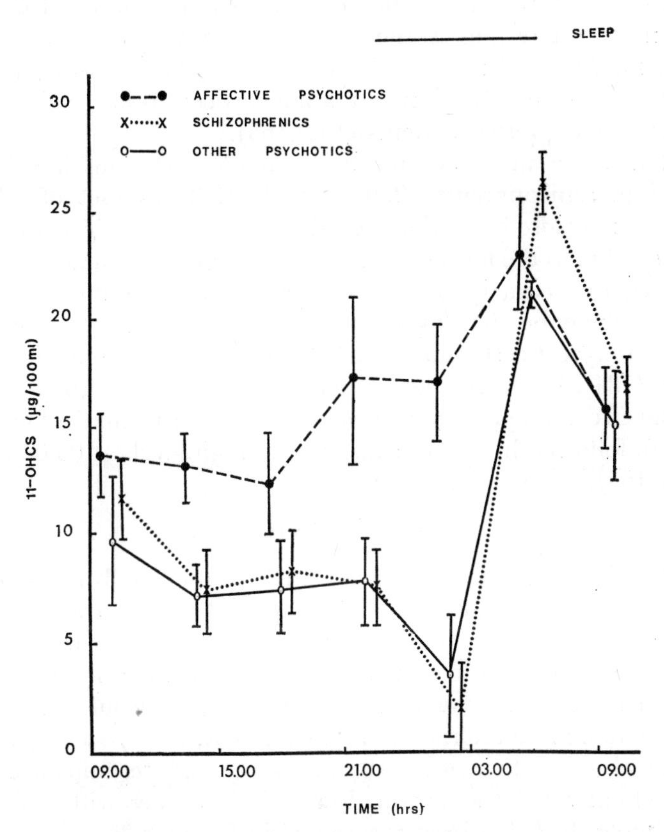

FIG. 11.2. Plasma corticosteroid concentrations in three groups of psychotic subjects. (Conroy, R. T. W. L., Hughes, B. D. and Mills, J. N., Fig. 1, 165.)

patients after recovery (100). Knapp *et al.* (480), taking 4-hourly samples, found significantly elevated 11-hydroxycorticosteroid levels at 04.00 hours as compared with a group of normal subjects. The authors by taking serial samples at 10.00 hours, 14.00 hours, 18.00 hours, 22.00 hours, 02.00 hours, 06.00 and again at 10.00 hours in three groups of psychiatric patients, were able to show that raised plasma corticosteroid values may be present throughout the 24 hours in affective psychotics, but are more readily observed in samples taken in the late evening or at night, and can even be observed in morning blood samples when these are compared with controls (Fig. 11.2). On clinical recovery, the plasma corticosteroid rhythm may be altered in affective psychotics (162, 206) but more interesting is the fact that despite gross psychiatric abnormalities and marked disturbances of the sleep/wakefulness cycle, the circadian rhythm of plasma corticosteroids persists, even in severely depressed patients (162, 165).

Changes in urine volume and circadian rhythm have been noted in schizophrenia (289, 404, 739). The onset of schizophrenia may also be marked by a disturbance of the sleep/wakefulness rhythm in the form of a prolonged period of wakefulness (78).

Hoagland *et al.*, 1953 (391) have suggested that adrenocortical secretion may be abnormal in schizophrenia, and elevated urinary 17-hydroxycorticosteroids have been found in acute schizophrenic states (784). We have noted supernormal plasma hydroxycorticosteroid circadian patterns in schizophrenics with a marked nocturnal fall followed by a very sharp rise to high levels at 06.00 hours (162, 165).

OBSTETRICS

The high incidence of normal birth in the small hours of the morning has been discussed in Chapter 8. By contrast, neither the onset nor termination of abnormal labours appears to have any definite relationship to clock time. The peak hour for stillbirths and births associated with neonatal mortality would, however, appear to be late afternoon (190, 448, 943).

The absence of circadian rhythm in the high serum concentrations of human chorionic gonadotrophin in women with hydatidiform mole (862) has been mentioned in Chapter 3.

MALIGNANCY

Circadian variations in mitotic frequency in human tumour cells have been claimed (854, 899). Statistical analysis of this data by the cosinor technique has led Garcia-Sainz and Halberg (272) to doubt some of these findings as regards squamous and basal cell carcinomata, but does suggest that a circadian rhythm in the mitotic activity of mammary carcinomata may exist (Fig. 11.3).

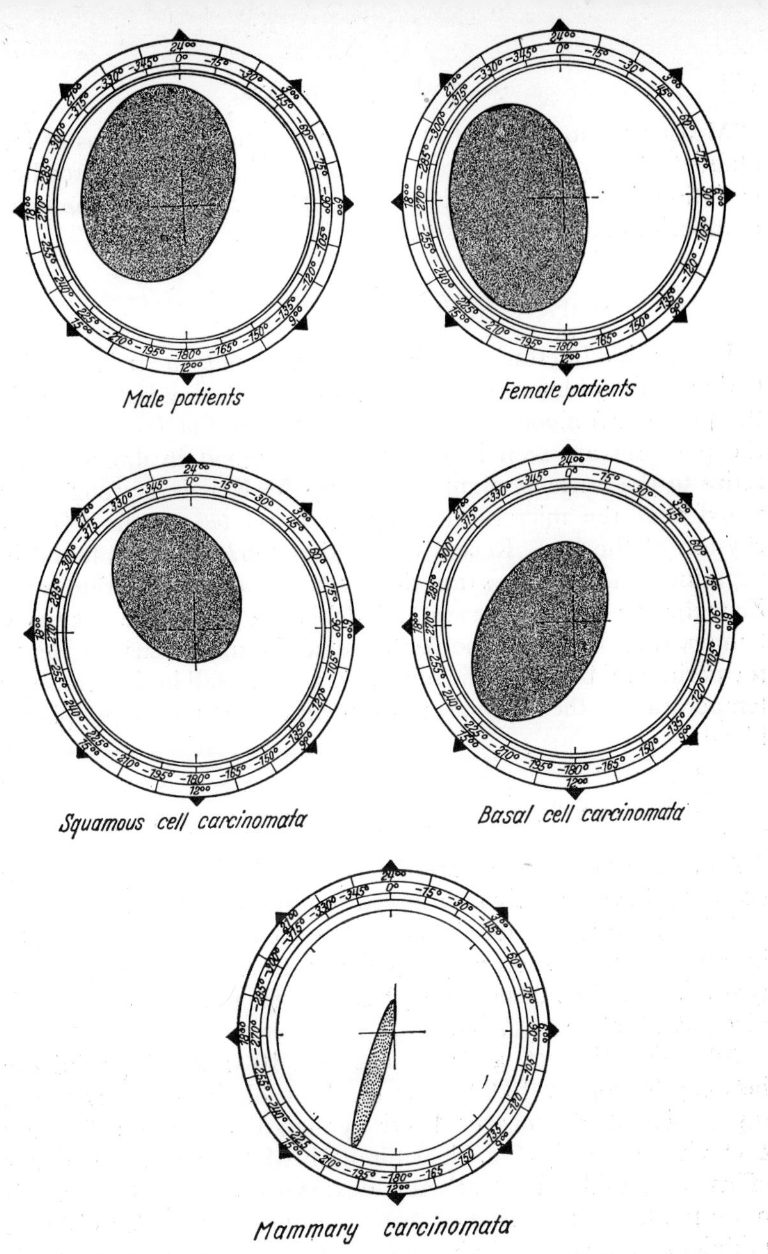

FIG. 11.3. Cosinor analysis of serial mitotic counts. In the cosinors summarizing counts from patients with squamous or basal cell carcinomata the error ellipses (shaded) cover the pole, i.e. a circadian rhythm is not indicated. In the cosinor derived from the mitotic counts in mammary carcinomata, the error ellipse does not cover the pole—a circadian rhythm is suggested. (Halberg, F., Tong, Y. L. and Johnson, E. A., Fig. 11, 347.)

Measurements of ^{32}P-orthophosphate uptake, by means of implanted Geiger counters, are reported to have shown circadian variations in radioactivity in malignant tissue in some of the small number of patients so far studied (858, 933).

PARASITIC INFESTATIONS

In parasitic infestation with *Wucheria bancrofti* the microfilariae gather in the pulmonary capillaries during the day but appear in the peripheral blood at night, whereas with *Loa Loa* infestation the parasites are seen in the peripheral blood in daytime and retire to the lungs at night (365, 366). McKenzie in 1882 (663) noted that the microfilarial rhythm is related to the 24-hour rhythm of the host. Recently Hawking (367, 368) has found in experiments on monkeys that the host's temperature rhythm is the Zeitgeber for *Wucheria bancrofti.* A deliberate rise in the nocturnal temperature of infested monkeys caused the microfilariae to leave the peripheral blood for the lungs and a forced fall in the monkey's temperature in the daytime caused them to re-enter the peripheral blood.

PHARMACOLOGY AND THERAPEUTICS

A number of circadian variations in the effects of drugs and their excretion have recently been noted in human subjects.

It has been found, for example, that the effects of an oral antihistaminic preparation may last for 15-17 hours when the drug is administered at 07.00 hours, but only for 6-8 hours when it is given at 19.00 hours (751).

Administration of a single dose of 10 mg of prednisone between 08.00 and 10.00 hours has been observed to cause less suppression of adrenocortical activity than divided doses of 2.5 mg each given 6-hourly (Fig. 11.4). Similarly Metopirone has a greater effect after midnight (571) and dexamethasone causes a much more marked suppression of corticosteroid output when given at midnight than when administered at 08.00 or 16.00 hours (673). Thus for complete suppression of ACTH released and adrenocortical activity, as in the treatment of the adrenogenital syndrome, dexamethasone should be given at midnight. Administration of flumethasone has been found (806) to cause a 40 per cent suppression of mean 17-hydroxycorticosteroid excretion when given at

08.00 as compared with 61 per cent at 16.00 hours and 83 per cent at 24.00 hours, and it has been suggested that the preferred method for administering glucocorticoids should be a single early morning dose given once daily or on alternate days (806). The

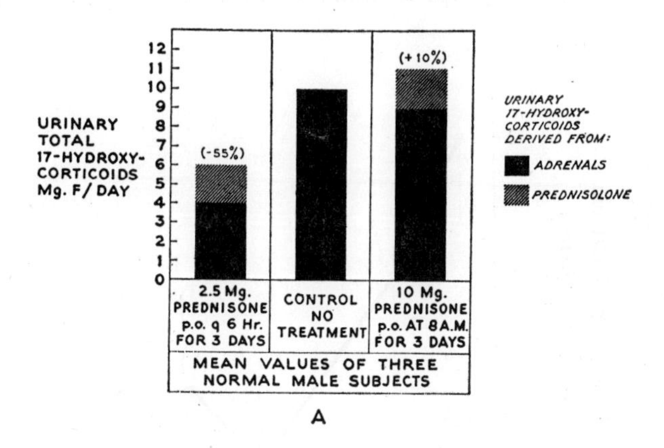

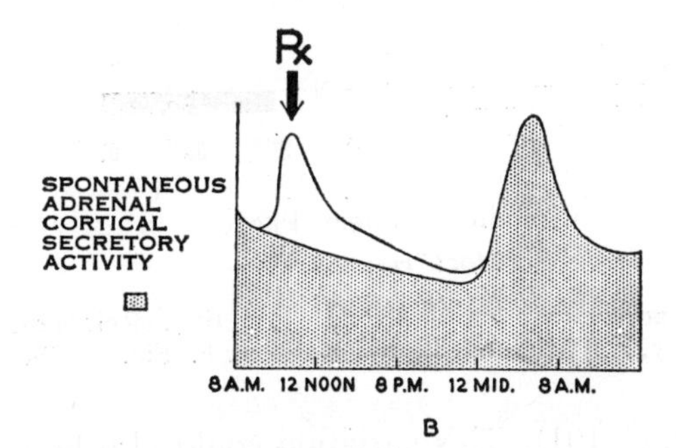

FIG. 11.4. Suppression of adrenocortical function when a single 10 mg dose of prednisone is given at 08.00 hours as compared with effect of four divided doses of 2.5 mg each at 6-hourly intervals. In lower figure shaded area represents endogenous steroid, unshaded exogenous steroid given at 08.00 hours. (DiRaimondo, V. C. and Forsham, P. H., Fig. 1, 196.)

succeeding circadian wave of ACTH secretion is then less likely to be blocked by high levels of circulating corticosteroids. A single dose also appears to be less associated with side effects such as peptic ulceration than divided doses given over the course of the day (187, 219, 357, 745, 836).

A circadian rhythm in the urinary excretion of amphetamine and methylamphetamine has been reported (52, 53, 54) and appears to be due to changes in urinary pH, as maintaining the pH at fairly constant values abolished the rhythm. Studies on the plasma concentrations of sulphonamides in man have shown a mean half life for sulphasymazine three times greater at night

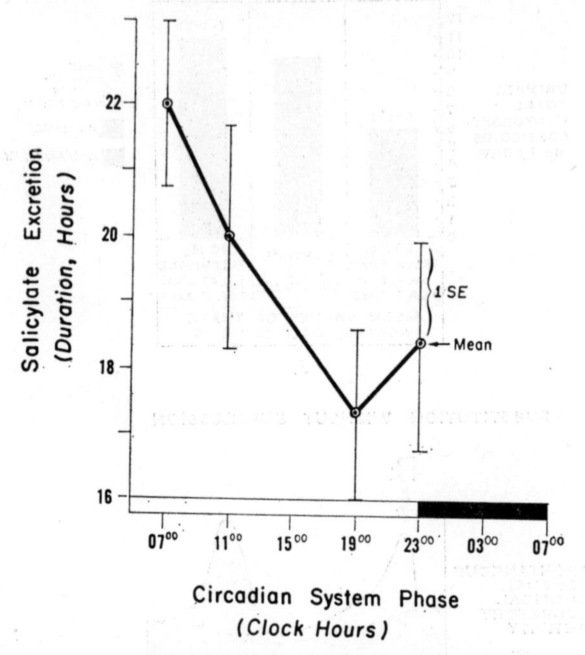

Fig. 11.5. Circadian variation in salicylate excretion. (Reinberg, A., Zagula-Mally, Z. W., Ghata, J. and Halberg, F., Fig. 1, 753.)

than during the day (191). This variation could also be due to changes in urinary pH from the low values seen during sleep to the high pH after awakening.

The duration of salicylate excretion in man has been shown to have a circadian rhythm (753). Salicylate excretion in six subjects given salicylates orally at 07.00, 11.00, 19.00 or 23.00 hours, was observed. When administration was at 07.00 hours the mean duration of excretion was 22 hours, with administration at 19.00 hours the mean duration of excretion was 17 hours. Intermediate values were seen for administration at the other times (Fig. 11.5).

Cardiac patients are stated to be several times more sensitive to digitalis during the night than during the day (830).

Patients on guanethidine, a long acting sympatholytic drug, exhibit marked and consistent circadian variations in arterial blood pressure (208). The morning pressure is low and hypotensive symptoms are common and may be disabling. The cause of this

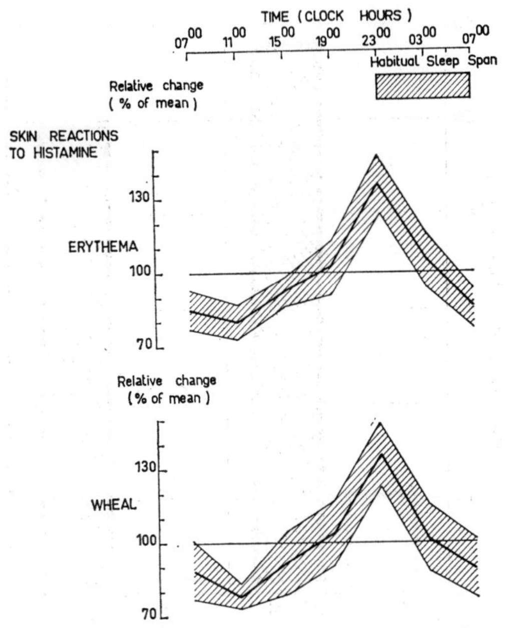

Fig. 11.6. Circadian variations in skin reactions to intradermal injections of histamine. Shaded areas represent $\pm$ 1 standard error of the mean. (Reinberg, A. and Sidi, E., Fig. 1, 751.)

daily rhythm in activity of guanethidine is not known but may be related to changes in plasma volume (169).

It is possible that in the near future much more information will become available on circadian rhythms in pharmacology and in therapeutics and this may well necessitate a reappraisal of traditional habits in prescribing and attention to the timing of drug administration.

MORTALITY AND SUSCEPTIBILITY

A circadian rhythm in allergic reactions has been described by Reinberg and his colleagues (746, 748, 752) who found that the cutaneous response to intradermal injections of histamine reaches a peak at about 23.00 hours (Fig. 11.6). A 24-hour variation in the effects of a bacterial pyrogen on the pituitary axis has also been noted (855).

The early morning peak in the incidence of death presumably results from the circadian variations in one or more physiological

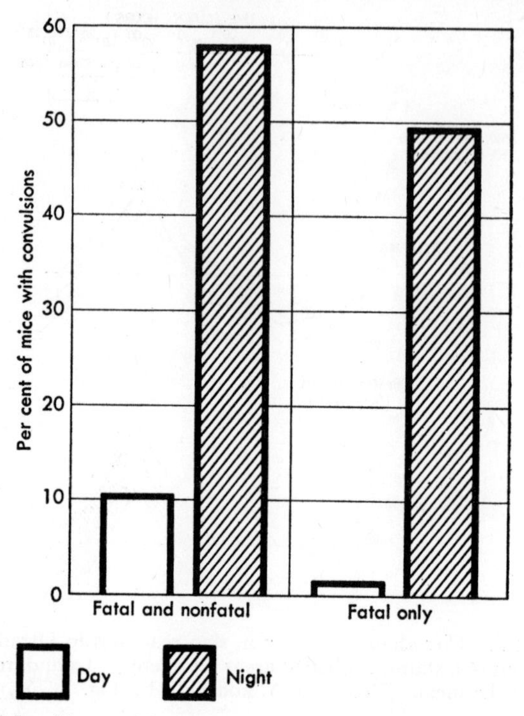

FIG. 11.7. Unequal incidence rate by day and by night of convulsions, and deaths from convulsions, in mice following exposure to auditory stimulation. (Halberg, F. and Howard, R. B., Fig. 3, 341.)

functions (267, 322, 427, 445) as does the similar incidence of post-operative death (267).

Mortality in mice from chemical or physical agent or toxins may be several times greater at one time of day than at another (239, 325, 326, 331, 342, 343, 648 and see Fig. 11.7). The therapeutic implications of these findings do not appear to have received much attention as yet, although they may well become of considerable clinical significance. It would be interesting to discover more about how human susceptibility and mortality may vary.

It will be seen from this account that even less is known about abnormal rhythms in disease than about normal circadian rhythms. In some conditions, the disturbance is probably a direct influence of the disease upon the function studied. For example, recumbency in itself causes increased urinary excretion of water and sodium, but in the healthy man the usual time of recumbency, the night, coincides with the time when the circadian influence causes water and sodium retention. In heart failure the postural effect is much greater, so that although the circadian influence may be operating in a normal manner, it is overridden by the postural effect. In other diseases, as of the adrenals, the disease process probably interferes with transmission between clock and hands, while some of the grosser diseases of the central nervous system may affect the clock itself. Yet other diseases are associated with disturbances of the rhythmic process which we cannot as yet interpret, including those in which rhythms of a period far from 24 hours become prominent. In the face of this ignorance of the fundamental facts and mechanisms of circadian rhythms in pathology and in pharmacology it is not surprising that their role in medicine and therapeutics is a field which is only just beginning to be explored.

REFERENCES

1 Aanonsen, A. (1959). Medical problems of shift work. *Industr. Med. Surg.* 28: 422-427.

2 Aanonsen, A. (1964). *Shift Work and Health.* Norwegian monographs on Medical Science, Oslo.

3 Abernethy, J. D., Maurizi, J. J. and Farhi, L. E. (1967). Diurnal variations in urinary-alveolar N_2 difference and effects of recumbency. *J. appl. Physiol.* 23: 875-879.

4 Acland, J. D. and Gould, A. H. (1956). Normal variation in the count of circulating eosinophils in man. *J. Physiol.* 133: 456-466.

5 Adam, J. M. and Ferres, H. M. (1954). Observations on oral and rectal temperature in the humid tropics and in a temperate climate. *J. Physiol.* 125: 21P.

6 Adam, J. M., Lobban, M. C. and Tredre, B. E. (1965). Diurnal rhythms of renal excretion and of body temperature in Indian subjects after a sudden change of environment. *J. Physiol.* 177: 18-19P.

7 Adams, O. S. and Chiles, W. D. (1961). Human performance as a function of the work-rest ratio during prolonged confinement. *USAF, ASD Technical Report*, No. 61-720.

8 Adkins, S. (1964). Performance, heart rate, and respiration rate on the day-night continuum. *Percept. Mot. Skills*, 18: 409-412.

9 Albertsen, K. and Petersen, V. P. (1968). Urinary rhythm after renal transplantation. *Lancet*, 2: 636-637.

10 Allbutt, T. C. (1872). The effect of exercise on the bodily temperature. *J. Anat., Lond.* 7: 106-119.

11 Alluisi, E. A. and Chiles, W. D. (1967). Sustained performance, work-rest scheduling, and diurnal rhythms in man. *Acta psychol., Hague*, 27: 436-442.

12 Alluisi, E. A., Chiles, W. D. and Hall, T. J. (1964). Combined effects of sleep loss and demanding work-rest schedules on crew performance. *AMRL Technical Report*, No. 64-63.

13 Alluisi, E. A., Chiles, W. D. and Hall, T. J. and Hawkes, G. R. (1963). Human group performance during confinement. *AMRL Technical Report*, No. 63-87.

14 Altman, K. A. and Stellate, R. (1963). Variation of protein content of urine in a 24-hour period. *Clin. Chem.* 9: 63-69.

15 Anderson, J. and Parsons, V. (1963). Tubular maximal resorptive rate for inorganic phosphate (TmP) in normal subjects. *Clin. Sci.* 25: 431-441.

16 Andres, R., Cader, G., Goldman, P. and Zierler, K. L. (1957). Net potassium movement between resting muscle and plasma in man in the basal state and during the night. *J. clin. Invest.* 36: 723-729.

17 Appel, W. (1938). Über die Tagesschwankungen der Eosinophilen. *Z. ges. exp. Med.* 104: 15-21.

18 Appel, W. and Hansen, K. J. (1952). Lichteinwirkung, Tagesrhythmik der eosinophilen Leukozyten und Hypophysennebennierenrindensystem. *Dtsch. Arch. Klin. Med.* 199: 530-537.

19 Aron, E., Thouvenot, J., Martin, A. and Groussin, P. (1966). Enregistrement prolongé de l'activité motrice du colon gauche. Étude des

variations nycthémérales sur 50 cas. *Arch. Franc. Mal. Appar. Dig.* 55: 495-506.

20 Aschoff, J. (1955). Der Tagesgang der Körpertemperatur beim Menschen. *Klin. Wschr.* 33: 545-551.

21 Aschoff, J. (1955). Exogene und endogene Komponente der 24 Stunden-periodik bei Tier und Mensch. *Naturwissenschaften.* 42: 569-575.

22 Aschoff, J. (1958). Tierische Periodik unter dem Einfluss von Zeitgebern. *Z. Tierpsychol.* 15: 1-30.

23 Aschoff, J. (1959). Periodik Licht—und dunkelaktiver Tiere unter konstanten Umgebungsbedingungen. *Pflüg. Arch. ges. Physiol.* 270: 9.

24 Aschoff, J. (1960). Exogenous and endogenous components in circadian rhythms. *Cold Spr. Harb. Symp. quant. Biol.* 25: 11-28.

25 Aschoff, J. (1960). Contribution to discussion in Lobban, M.C. The entrainment of circadian rhythms in man. *Cold Spr. Harb. Symp. quant. Biol.* 25: 325-332.

26 Aschoff, J. (1963). Diurnal rhythms. *Ann. Rev. Physiol.* 25: 581-600.

27 Aschoff, J. (1965). *Circadian Clocks.* Proceedings of the Feldafing Summer School. Amsterdam: North Holland.

28 Aschoff, J. (1967). Die minimale Wärmedurchgangszahl des Menschen am Tage und in der Nacht. *Pflüg. Arch. ges. Physiol.* 295: 184-196.

29 Aschoff, J., Gerecke, U. and Wever, R. (1967). Phasenbeziehungen zwischen den circadianen Perioden der Aktivität und der Kerntemperatur beim Menschen. *Pflüg. Arch. ges. Physiol.* 295: 173-183.

30 Aschoff, J., Gerecke, U. and Wever, R. (1967). Desynchronisation of human circadian rhythms. *Jap. J. Physiol.* 17: 450-457.

31 Aschoff, J., Pöppel, E. and Wever, R. (1969). Circadiane Periodik des Menschen unter dem Einfluss von Licht-Dunkel-Wechseln unterschiedlicher Periode. *Pflüg. Arch. ges. Physiol.* 306: 58-70.

32 Aschoff, J. and Wever, R. (1962). Spontanperiodik des Menschen bei Ausschluss aller Zeitgeber. *Naturwissenschaften*, 49: 337-342.

33 Baddeley, A. D. (1966). Time estimation at reduced body temperature. *Amer. J. Psychol.* 79: 475-479.

34 Baechler, C., Borth, R. and Menzi, A. (1966). Immunoassay of human chorionic gonadotrophin: Diurnal variation in urinary antigen excretion. *Acta. endocr., Copenhagen.* 52: 199-209.

35 Baerensprung, F. (1851). Untersuchungen über die Temperaturverhältnisse des Foetus und des erwachsenen Menschen im gesunden und kranken Zustande. *Arch. f. anat. physiol. u wissensch. med.* 2: 126-175.

36 Baerensprung, F. (1852). Untersuchungen über die Temperaturverhältnisse des Menschen im gesunden und kranken Zustande. *Arch. Anat. Physiol. u wissensch. med.* 3: 217-286.

37 Bain, W. E. S. (1954). Variations in the episcleral venous pressure in relation to glaucoma. *Brit. J. Ophthal.* 38: 129-135.

38 Balzer, E. (1953). 24-Stunden-Rhythmus des Serumbilirubins. *Acta. med. scand. Supp.* 278: 67-70.

39 Banks, O. (1956). Continuous shift work: the attitudes of wives. *Occup. Psychol.* 30: 69-84.

40 Bardswell, N. D. and Chapman, J. E. (1911). Some observations upon

the deep temperature of the human body at rest and after muscular exertion. *Brit. med. J.* 1: 1106-1110.

41 Barker, E. S., Singer, R. B., Elkinton, J. R. and Clark, J. K. (1957). The renal response in man to acute experimental respiratory alkalosis and acidosis. *J. clin. Invest.* 36: 515-529.

42 Barnett, G. D. and Blume, F. E. (1938). Alkaline tides. *J. clin. Invest.* 17: 159-165.

43 Barraclough, M. A., Jones, N. F. and Marsden, C. D. (1967). Effect of angiotensin on renal function in rat. *Amer. J. Physiol.* 212: 1153-1158.

44 Barraclough, M. A. and Mills, I. H. (1965). Bradykinin and renal function. *Clin. Sci.* 28: 69-74.

45 Bartter, F. C. (1964). In *Aldosterone*: a symposium organised by the Council for International Organizations of Medical Sciences. Baulieu, E. E. and Robel, P. (eds.). Oxford: Blackwell.

46 Bartter, F. C., Biglieri, E. G., Pronove, P. & Delea, C. S. (1958). Effect of changes in intravascular volume on aldosterone secretion in man. In *Aldosterone*: Muller, A. F. and O'Connor, C. (eds.). London: J. & A. Churchill Ltd.

47 Bartter, F. C., Delea, C. S. and Halberg, F, (1962). A map of blood and urinary changes related to circadian variations in adrenal cortical functions in normal subjects. *Ann. N.Y. Acad. Sci.* 98: 969-983.

48 Bartter, F. C. and Fourman, P. (1962). The different effects of aldosterone-like steroids and hydrocortisone-like steroids on urinary excretion of potassium and acid. *Metabolism*, 11: 6-20.

49 Bayliss, R. I. S. (1955). Factors influencing adrenocortical activity in health and disease. *Brit. med. J.* 1: 495-501.

50 Bayliss, R. I. S. (1955). Contribution to discussion in: Elmadjian, F. Adrenocortical function of combat infantrymen in Korea. In *The Human Adrenal Cortex*. Wolstenholme, G. E. W., Cameron, M .P. and Etherington, J. (eds.). *Ciba Found. Colloq. Endocrin.* 8: 627-653.

51 Bazett, H. C., Thurlow, S., Crowell, C. and Stewart, W. (1924). Studies on the effects of baths on man. II. The diuresis caused by warm baths, together with some observations on urinary tides. *Amer. J. Physiol.* 70: 430-452.

52 Beckett, A. H. and Rowland, M. (1964). Rhythmic urinary excretion of amphetamine in man. *Nature, Lond.* 204: 1203-1204.

53 Beckett, A. H. and Rowland, M. (1965). Urinary excretion of methylamphetamine in man. *Nature, Lond.* 206: 1260-1261.

54 Beckett, A. H., Rowland, M. and Turner, P. (1965). Influence of urinary pH on excretion of amphetamine. *Lancet*, 1: 303.

55 Bell, C. R. (1965). Time estimation and increases in body temperature. *J. exp. Psychol.* 70: 232-234.

56 Bell, C. R. and Provins, K. A. (1963). Relation between physiological responses to environmental heat and time judgments. *J. exp. Psychol.* 66: 572-579.

57 Belluc, S., Chaussin, J., Laugier, H. and Ranson, T. (1938). Les variations nycthémérales dans l'élimination des principales substances de l'urine. *C. R. Acad. Sci., Paris.* 207: 90-92.

58 Benedict, F. G. (1904). Studies in body temperature. *Amer. J. Physiol.* 11: 145-169.
59 Benedict, F. G. and Snell, J. F. (1902). Körpertemperatur-Schwankungen mit besonderer Rücksicht auf den Einfluss, welchen die Umkehrung der täglichen Lebensgewohnheit beim Menschen ausübt. *Pflüg. Arch. ges. Physiol.* 90: 33-72.
60 Bercel, N. A. (1964). The periodic features of some seizure states. *Ann. N.Y. Acad. Sci.* 117: 555-562.
61 Berendes, H. W., Marte, E., Ertel, R. J., McCarthy, J. A., Anderson, J. A. and Halberg, F. (1960). Circadian physiologic rhythms and lowered blood 5-hydroxytryptamine in human subjects with defective mentality. *Physiologist*, 3: 20.
62 Beringer, A. (1955). Über die rhythmischen Schwankungen im Fett-und Glykogengehalt der Leber, ihre Beeinflussing sowie ihre Ursache. *Acta. med. scand. Supp.* 307: 172-174.
63 Berliner, R. W. (1960). Renal mechanisms for potassium excretion. *Harvey Lect.* 55: 141-171.
64 Berlyne, G. M., Mallick, N. P., Seedat, Y. K., Edwards, E. C., Harris, R. and Orr, W. McN. (1968). Abnormal urinary rhythm after renal transplantation in man. *Lancet*, 2: 435-436.
65 Berry, C. A. (1969). Preliminary clinical report of the medical aspects of Apollos 7 and 8. *NASA Technical Memorandum* X-58027.
66 Besterman, E., Myat, G. and Travadi, V. (1967). Diurnal variations of platelet stickiness compared with effects produced by adrenaline. *Brit. med. J.* 1: 597-600.
67 Bevan, A. T., Honour, A. J. and Stott, F. H. (1969). Direct arterial pressure recording in unrestricted man. *Clin. Sci.* 36: 329-344.
68 Billimoria, J. D., Drysdale, J., James, D. C. O. and MacLagan, N. F. (1959). Determination of fibrinolytic activity of whole blood with special reference to the effects of exercise and fat feeding. *Lancet*, 2: 471-475.
69 Birkenhager, W.H., Hellendoorn, H. B. A., and Gerbrandy, J. (1959). Effects of intravenous injections of calcium laevulinate on calcium and phosphate metabolism. *Clin. Sci.* 18: 45-53.
70 Bissada, N. F., Schaffer, E. M. and Haus, E. (1967). Circadian periodicity of human crevicular fluid flow. *J. Periodont.* 38: 36-40.
71 Bjerner, B. (1949). Alpha depression and lowered pulse rate during delayed actions in a serial reaction test. A study in sleep deprivation. *Acta. physiol. Scand.* 19, *Suppl.* 65: 4-93.
72 Bjerner, B., Holm, A. and Swennsson, A. (1955). Diurnal variation in mental performance. A study of three shift-workers. *Brit. J. industr. Med.* 12: 103-110.
73 Bjerner, B. and Swennsson, A. (1953). Shiftwork and rhythm. *Acta. med. scand. Supp.* 278: 102-107.
74 Blake, M. J. F. (1965). Physiological and temperamental correlates of performance at different times of day. *Ergonomics*, 8: 375-376.
75 Blake, M. J. F. (1967). Relationship between circadian rhythm of body temperature and introversion-extroversion. *Nature, Lond.* 215: 896-897.
76 Blake, M. J. F. (1967). Time of day effects on performance in a range of tasks. *Psychol. Sci.* 9: 349-350.

77 Blake, M. J. F. and Colquhoun, W. P. (1965). Experimental studies of shift work. *Ergonomics*, 8: 376.

78 Bliss, E. L., Clark, L. D. and West, C. D. (1959). Studies of sleep deprivation—relationship to schizophrenia. *Arch. Neurol. Psychiat., Chicago*, 81: 348-359.

79 Bliss, E. L., Sandberg, A. A., Nelson, D. H. and Eik-Nes, K. (1953). The normal levels of 17-hydroxycorticosteroids in the peripheral blood in man. *J. clin. Invest.* 32: 818-823.

80 Bloch, M. (1964). Rhythmic diurnal variation in limb blood flow in man. *Nature, Lond.* 202: 398-399.

81 Blum, A. S., Greenspan, F. S. and Magnum, J. (1968). Circadian rhythm of serum TSH in normal subjects. Proc. 3rd Int. Cong. of Endocrinology. *Excerpta med. Int. Cong. Series*, 157: 14.

82 Board, F., Persky, H. and Hamburg, D. A. (1956). Psychological stress and endocrine functions: Blood levels of adrenocortical and thyroid hormones in acutely disturbed patients. *Psychosom. Med.* 18: 324-333.

83 Board, F., Wadeson, R. and Persky, H. (1957). Depressive affect and endocrine functions. *Arch. Neurol. Psychiat., Chicago*, 78: 612-620.

84 Board of Trade (Great Britain) (1967). *Flight time limitations*, the evidence of excessive fatigue in aircrews. London: H.M.S.O.

85 Bock, H. E., von Oldershausen, H. F. and Tellesz, A. (1953). Was leistet die Leberpunktion bei der Tuberkulose? *Verh. dtsch. Ges. inn. Med.* 59: 351-361.

86 Bohm, E., Gernant, G. and Holmgren, H. (1940). Rytmiska variationer i blodets halt av chylomicron. *Nord. Med.* 8: 2389-2392.

87 Bonjer, F. H. (1960). Physiological aspects of shiftwork. *Proc. 13th Int. Congr. occup. Health*, 848-849.

88 Bonner, C. D. (1952). Eosinophil levels as index of adrenal responsiveness. Factors that affect value of eosinophil counts. *J. Amer. med. Ass.* 148: 634-637.

89 Bornstein, A. and Völker, H. (1926). Über die Schwankungen des Grundumsatzes. *Z. ges. exp. Med.* 53: 439-450.

90 Borst, J. G. G. and DeVries, L. A. (1950). The three types of 'natural' diuresis. *Lancet*, 2: 1-6.

91 Borth, R., Lunenfeld, B. and de Watteville, H. (1958). Lack of diurnal variation of chorionic gonadotrophin levels in pregnancy serum. *Acta. endocr., Copenhagen*, 29: 531-536.

92 Bowen, A. J. and Reeves, R. L. (1967). Diurnal variation in glucose tolerance. *Arch. intern. Med.* 199: 261-264.

93 Bowie, E. J., Tauxe, W. N., Sjoberg, W. E. Jr. and Yamaguchi, M. Y. (1963). Daily variation in the concentration of iron in serum. *Amer. J. clin. Path.* 40: 491-494.

94 Boyd, E. M. (1935). Diurnal variations in plasma lipids. *J. biol. Chem.* 110: 61-70.

95 Boyd, T. A., Hassard, D. T., Patrick, A. and McLeod, L. E. (1961). The relation of diurnal variation of plasma corticoid levels and intraocular pressure in glaucoma. *Trans. Canad. ophthal. Soc.* 24: 119-134.

96 Boyd, T. A. and McLeod, L. E. (1964). Circadian rhythms of plasma

corticoid levels, intraocular pressure and aqueous outflow facility in normal and glaucomatous eyes. *Ann. N.Y. Acad. Sci.* 117: 597-613.

97 BOYD, T. A., MCLEOD, L. E., HASSARD, D. T. and PATRICK, A. (1962). Relation of diurnal variation of plasma corticoid levels and intraocular pressure in glaucoma. *Canad. med. Ass. J.* 86: 772-775.

98 BRANWOOD, A. W. (1946). Post-prandial variation in haemoglobin. *Edinb. med. J.* 53: 125-133.

99 BRAZEAU, P. and GILMAN, A. (1953). Effect of plasma CO_2 tension on renal tubular reabsorption of bicarbonate. *Amer. J. Physiol.* 175: 33-38.

100 BRIDGES, P. K. and JONES, M. T. (1966). The diurnal rhythm of plasma cortisol concentration in depression. *Brit. J. Psychiat.* 112: 1257-1261.

101 BRINDLEY, G. S. (1954). Intrinsic 24-hour rhythms in human physiology, and their relevance to the planning of working programmes. *R.A.F. Inst. Aviation FPRC.* 871.

102 BRISCOE, A. M. and RAGAN, C. (1966). Diurnal variations in calcium and magnesium excretion in man. *Metabolism,* 15: 1002-1010.

103 BROADBENT, D. E. (1958). *Perception and Communication.* London, New York, Paris and Los Angeles: Pergamon Press.

104 BROADHURST, H. C. and LEATHES, J. B. (1920). The excretion of phosphoric acid in the urine. *J. Physiol.* 54: 28-29P.

105 BROCH, O. J. and HAUGEN, H. N. (1950). The effects of adrenaline on the number of circulating eosinophils and on the excretion of uric acid and creatinine. *Acta. Endocr., Copenhagen.* 5: 143-150.

106 BROD, J. and FENCL, V. (1957). Diurnal variations of systemic and renal haemodynamics in normal subjects and in hypertensive diseases. *Cardiologia, Basle,* 31: 494-499.

107 BRODERS, A. C. and DUBLIN, W. B. (1939). Rhythmicity of mitosis in epidermis of human beings. *Proc. Mayo Clin.* 19: 423-425.

108 BROOKS, H. and CARROLL, J. H. (1912). A clinical study of the effects of sleep and rest on blood pressure. *Arch. intern. Med.* 10: 97-102.

109 BROOKS, R. V., DUPRE, J., GOGATE, A. N., MILLS, I. H. and PRUNTY, F. T. G. (1963). Appraisal of adrenocortical hyperfunction: Patients with Cushing's Syndrome or 'non-endocrine' tumours. *J. clin. Endocr.* 23: 725-736.

110 BROOKSBANK, B. W. L. and COPPEN, A. (1967). Plasma 11-hydroxycorticosteroids in affective disorders. *Brit. J. Psychiat.* 113: 395-404.

111 BRORSON, I. (1964). Concentration of corticosterone and hydrocortisone in the plasma of patients with Cushing's Syndrome caused by hyperplasia or tumours of the adrenal cortex. *Acta. chir. scand.* 127: 162-171.

112 BROWN, F. A. Jr. (1959). Living clocks. *Science,* 130: 1535-1544.

113 BROWN, F. A., Jr. (1960). Response to pervasive geophysical factors and the biological clock problem. *Cold Spr. Harb. Symp. quant. Biol.* 25: 57-71.

114 BROWN, F. A., Jr. (1961). Extrinsic Rhythmicality and the Timing of the Circadian Rhythms. In *Circadian Systems.* Report 39th Ross Conference on Pediatric Research, S. J. Fomon (ed.), Columbus, Ohio: Ross Labora: tories, 28-31.

115 BROWN, F. A., Jr. (1962). Extrinsic rhythmicality: a reference frame for biological rhythms under so-called constant conditions. *Ann. N.Y. Acad. Sci.* 98: 775-787.

116 Brown, F. A., Jr., Shriner, J. and Ralph, C. L. (1956). Solar and lunar rhythmicity in the rat in 'constant conditions' and the mechanism of physiological time measurement. *Amer. J. Physiol.* 184: 491-496.
117 Brown, F. A., Jr., Webb, H. M. and Bennett, M. F. (1958). Comparisons of some fluctuations in cosmic radiation and in organismic activity during 1954, 1955 and 1956. *Amer. J. Physiol.* 195: 237-243.
118 Brown, H., Englert, E., Wallach, S. and Simons, E. L. (1957). Metabolism of free and conjugated 17-hydroxycorticosteroids in normal subjects. *J. clin. Endocr.* 17: 1191-1201.
119 Brown, J. J., Davies, D. L., Lever, A. F. and Robertson, J. I. S. (1966). Variation of plasma renin concentration during a 24-hour period. *J. Endocrin.* 34: 129-130.
120 Brown, J. M. and Berry, R. J. (1968). The relationship between diurnal variation of the number of cells in mitosis and of the number of cells synthesizing DNA in the epithelium of the hamster cheek pouch. *Cell Tissue Kinet.* 1: 23-33.
121 Browne, R. C. (1949). The day and night performance of teleprinter switchboard operators. *Occup. Psychol.* 23: 121-126.
122 Browne, R. C. (1961). The day and night performance rhythm in industry. Report 5*th Conf. Soc. Biol. Rhythm.* Stockholm: Aco Print, 61-64.
123 Bruger, M. and Somach, I. (1932). The diurnal variations of the cholesterol content of the blood. *J. biol. Chem.* 97: 23-30.
124 Brunton, C. E. (1933). The acid output of the kidney and the so-called alkaline tide. *Physiol. Rev.* 13: 372-399.
125 Brüschke, G. and Volkheimer, G. (1956). Untersuchungen zur Frage tagesrhythmischer Schwankungen des Serumbilirubinspiegels. *Z. ges. inn. Med.* 11: 804-806.
126 Brush, C. E. and Fayerweather, R. (1901). Observations on the changes in blood pressure during normal sleep. *Amer. J. Physiol.* 5: 199-210.
127 Bruusgaard, A. (1950). An investigation of inquiry on health problems in shift-works in the Norwegian paper industry. Norwegian State Factory Inspectorate, Oslo, 1949. *A.M. Arch. industr. Hyg.* 2: 465.
128 Buckell, M. and Elliott, F. A. (1959). Diurnal fluctuation of plasma fibrinolytic activity in normal males. *Lancet*, 1: 660-662.
129 Bullough, W. S. and Laurence, E. B. (1968). The role of glucocorticoid hormones in the control of epidermal mitosis. *Cell Tissue Kinet.* 1: 5-10.
130 Bülow, K. (1963). Respiration and wakefulness in man. *Acta physiol. scand.* 59, *suppl.* 209: 1-110.
131 Bünning, E. (1964). *The Physiological Clock.* Heidelberg: Springer-Verlag.
132 Burckard, E. and Kayser, C. (1947). L'inversion du rythme nycthéméral de la température chez l'homme. *C.R. Soc. Biol., Paris,* 141: 1265-1268.
133 Burton, A. C. (1956). The clinical importance of the physiology of temperature regulation. *Canad. med. Ass. J.* 75: 715-720.
134 Cade, R., Shires, D. L., Barrow, M. V. and Thomas, W. C. Jr. (1967). Abnormal diurnal variation of plasma cortisol in patients with renovascular hypertension. *J. clin. Endocrin.* 27: 800-806.
135 Campbell, J. A. and Webster, T. A. (1921). Day and night urine during complete rest, laboratory routine, light muscular work, and oxygen administration. *Biochem. J.* 15: 660-664.

136 CARRUTHERS, B. M., COPP, D. H. and MCINTOSH, H. W. (1964). Diurnal variation in urinary excretion of calcium and phosphate and its relation to blood levels. *J. Lab. clin. Med.* 63: 959-968.

137 CHADWICK-JONES, J. (1967). Shift-working: Physiological effects and social behaviour. *Brit. J. industr. Relations*, 5: 237-243.

138 CHANCE, B., SCHOENER, B. and ELSAESSER, S. (1964). Control of the waveform of oscillations of the reduced pyridine nucleotide level in a cell-free extract. *Proc. nat. Acad. Sci., Wash.* 52; 337-341.

139 CHANDLER, H. L., LAWRY, E. Y., POTEE, K. G. and MANN, G. V. (1953). Spontaneous and induced variations in serum lipoproteins. *Circulation*, 8: 723-731.

140 CHARLES, E. (1953). The hour of birth, a study of the distribution of times of onset of labour and of delivery throughout the 24-hour period. *Brit. J. prev. soc. Med.* 7: 43-59.

141 CHILES, W. D., ALLUISI, E. A. and ADAMS, O. S. (1968). Work schedules and performance during confinement. *Hum. Factors.* 10: 143-196.

142 CHOVNICK, A. (1960). Biological Clocks. *Cold Spr. Harb. Symp. quant. Biol.* 25.

143 CINKOTAI, F. F. and THOMPSON, M. L. (1966). Diurnal variation in pulmonary diffusing capacity for carbon monoxide. *J. appl. Physiol.* 21: 539-541.

144 CLARK, R. H. and BAKER, B. L. (1964). Circadian periodicity in the concentration of prolactin in the rat hypophysis. *Science*, 143: 375-376.

145 CLAYTON, G. W., LIBRIK, L., GARDNER, R. L. and GUILLEMIN, R. (1963). Studies on the circadian rhythm of pituitary adrenocorticotropic release in man. *J. clin. Endocrin.* 23: 975-980.

146 CLEGG, B. R. and SCHAEFER, K. E. (1966). Studies of circadian cycles in human subjects during prolonged isolation in a constant environment using eight channel telemetry systems. *SMRL Memo Report*, 66-4. U.S.N. Submarine Medical Center, Groton, Connecticut.

147 CLEGG, B. R. and SCHAEFER, K. E. (1966). Measurement of periodicity and phase shift of physiological functions in isolation experiments. *Aerospace Med.* 37: 271.

148 CLOUDSLEY-THOMPSON, J. L. (1961). *Rhythmic activity in animal physiology and behaviour.* New York and London: Academic Press.

149 COHEN, I. and DODDS, E. C. (1924). Twenty-four hour observations on the metabolism of normal and starving subjects. *J. Physiol.* 59: 259-270.

150 COHEN, J. (1967). *Psychological Time in Health and Disease.* Springfield, Illinois: Charles C. Thomas.

151 COHEN, M., STIEFEL, M., REDDY, W. J. and LAIDLAW, J. C. (1958). The secretion and disposition of cortisol during pregnancy. *J. clin. Endocrin.* 18: 1076-1092.

152 COLE, L. C. (1957). Biological clock in the unicorn. *Science*, 125: 874-876.

153 COLIN, J., TIMBAL, J., BOUTELIER, C., HOUDAS, Y. and SIFFRE, M. (1968). Rhythm of the rectal temperature during a 6-month free-running experiment. *J. appl. Physiol.* 25: 170-176.

154 COLQUHOUN, W. P., (1960). Temperament, inspection efficiency and time of day. *Ergonomics*, 3: 377-378.

155 COLQUHOUN, W. P., BLAKE, M. J. F. and EDWARDS, R. S. (1968). Experi-

mental studies of shift-work. I. A comparison of 'rotating' and 'stabilized' 4-hour shift systems. *Ergonomics*, 11: 437-453.

156 Colquhoun, W. P., Blake, M. J. F. and Edwards, R. S. (1968). Experimental studies of shift-work. II: Stabilized 8-hour shift systems. *Ergonomics*, 11: 527-546.

157 Conroy, R. T. W. L. (1966). Plasma levels of adrenal corticosteroids after prolonged isolation. *J. Physiol.* 189: 31-32P.

158 Conroy, R. T. W. L. (1967). Circadian rhythm of 11-hydroxycorticosteroids in night workers. *J. Physiol.* 191: 21-22P.

159 Conroy, R. T. W. L. (1967). Circadian rhythm of plasma corticosteroids during the imposition of a 12-hour time schedule. *J. Physiol.* 194: 19-20P.

160 Conroy, R. T. W. L. (1969). Circadian rhythms. *J. Roy. Coll. Surg. Irl.* 5: 43-54.

161 Conroy, R. T. W. L., Elliott, A. L., Fort, A. and Mills, J. N. (1969). Circadian rhythms before and after a flight from India. *J. Physiol.* 204: 85P.

162 Conroy, R. T. W. L., Elliott, A. L., Hughes, B. D. and Mills, J. N. (1968). Physiological variables in some psychiatric patients. *J. Physiol.* 196: 129-130P.

163 Conroy, R. T. W. L., Elliott, A. L. and Mills, J. N. (1968). Adaptation of physiological variables to time zone transitions. *J. Physiol.* 197: 84-85P.

164 Conroy, R. T. W. L. and Hall, M. D. (1969). Adrenal cortical function and body temperature rhythms after a transatlantic flight. *J. Physiol.* 200: 123P.

165 Conroy, R. T. W. L., Hughes, B. D. and Mills, J. N. (1968). Circadian rhythm of plasma 11-hydroxycorticosteroids in psychiatric disorders. *Brit. Med. J.* 3: 504-507.

166 Cooper, Z. K. (1939). Mitotic rhythm in human epidermis. *J. invest. Derm.* 2: 289-300.

167 Cooper, Z. K. and Schiff, A. (1938). Mitotic rhythm in human epidermis. *Proc. Soc. exp. Biol., N.Y.* 39: 323-324.

168 Crabbé, J., Ross, E. J. and Thorn, G. W. (1958). The significance of the secretion of aldosterone during dietary sodium deprivation in normal subjects. *J. clin. Endocrin.* 18: 1159-1177.

169 Cranston, W. I. and Brown, W. (1963). Diurnal variation in plasma volume in normal and hypertensive subjects. *Clin. Sci.* 25: 107-114.

170 Crawford, M. L. J. and Thor, D. H. (1964). Circadian activity and noise comparisons of two confined groups with and without reference to clock time. *Percept. Mot. Skills*, 19: 211-216.

171 Crockford, G. W. and Davies, C. T. M. (1969). Circadian variations in responses to submaximal exercise on a bicycle ergometer. *J. Physiol.* 201: 94-95P,

172 Crombie, A. (1873). The daily range of normal temperature in India with practical remarks on clinical thermometry. *Indian Ann. Med. Sci. Calcutta.* 16: 550-606.

173 Cullis, W. C., Oppenheimer, E. M. and Ross-Johnston, M. (1922). Temperature and other changes in women during the menstrual cycle. *Lancet*, 2: 954-956.

174 DAMROSCH, L. (1853). Über die täglichen Schwankungen der menschlichen Eigenwärme in normalen Zustande. *Dtsch. Klin.* 5: 317-319.

175 DANZ, C. F. and FUCHS, C. F. (1848). Physischmedicinische Topographie des Krieses Schmalkalden. *Schriften der Gesellschaft zur Beförderung des Gesammten Naturwissenschaften zu Marburg*, 6: 205-206.

176 DAVANGER, M. (1964). Diurnal variations of ocular pressure in normal and in glaucomatous eyes. *Acta. ophthal., Kbh.* 42: 764-772.

177 DAVIS, J. O., HOLMAN, J. E., CARPENTER, C. C. J., URQUHART, J. and HIGGINS, J. T. (1964). An extra-adrenal factor essential for chronic renal sodium retention in presence of increased sodium-retaining hormone. *Circulation Res.* 14: 17-31.

178 DAVY, J. (1844). Miscellaneous observations on animal heat. *Phil. Trans.* 1: 57-64.

179 DAVY, J. (1845). On the temperature of man. *Phil. Trans.* 2: 319-333.

180 DE AJURIAGUERRA, J. (Ed.) (1968). *Cycles Biologiques et Psychiatrie.* Symposium Bel-Air III. Geneva: Georg.

181 DE BONO, E. and MILLS, I. H. (1965). Intrarenal monitoring of cardiac output in the regulation of sodium excretion. *Lancet*, 2: 1027-1032.

182 DEIGHTON, T. (1933). Physical factors in body temperature maintenance and heat elimination. *Physiol. Rev.* 13: 427-465.

183 DE LA MARE, G. and SHIMMIN, S. (1964). Preferred patterns of duty in a flexible shift-working situation. *Occup. Psychol.* 38: 203-214.

184 DE LA MARE, G. and WALKER, J. (1965). Shift working: the arrangement of hours on night work. *Nature, Lond.* 208: 1127-1128.

185 DELL'ACQUA, G., JORES, A., CANIGGIA, A. and SOLLBERGER, A. (1962). *Atti VII Conf. Intern. Soc. Stud. Ritm. Biol.* Turin: Panminerva Medica.

186 DE MOOR, P., HEIRWEGH, K., HEREMANS, J. F. and DECLERCK-RASKIN, M. (1962). Protein binding of corticoids studied by gel filtration. *J. clin. Invest.* 41: 816-827.

187 DEMOS, C. H., KRASNER, F. and GROEL, J. T. (1964). A modified (once a day) corticosteroid dosage regime. *Clin. Pharm. Ther.* 5: 721-727.

188 DEMURA, H., WEST, C. D., NUGENT, C. A., NAKAGAWA, K. and TYLER, F. H. (1966). A sensitive radioimmunoassay for plasma ACTH levels. *J. clin. Endocrin.* 26: 1297-1302.

189 DENNY-BROWN, D. and ROBERTSON, E. G. (1934). Observations on records of local epileptic convulsions. *J. Neurol. Psychopath.* 15: 97-136.

190 DE PORTE, J. V. (1932). The prevalent hour of stillbirth. *Amer. J. Obstet. Gynec.* 23: 31-37.

191 DETTLI, L. and SPRING, P. (1967). Diurnal variations in the elimination rate of a sulfonamide in man. *Helv. med. Acta*, 33: 291-306.

192 DE VENECIA, G. and DAVIS, M. D. (1963). Diurnal variation of intraocular pressure in the normal eye. *Arch. Ophthal. Chicago*, 69: 752-757.

193 DE VRIES, L. A., TEN HOLT, S. P., VAN DAATSELAAF, J. J., MULDER, A., and BORST, J. G. G. (1960). Characteristic renal excretion patterns in response to physiological, pathological and pharmacological stimuli. *Clin. Chem. Acta.* 5: 915-937.

194 DE WARDENER, H. E., MILLS, I. H., CLAPHAM, W. F. and HAYTER, C. J. (1961). Studies on the efferent mechanism of the sodium diuresis which

follows the administration of intravenous saline in the dog. *Clin. Sci.* 21: 249-258.

195 Dickinson, C. J. (1965). *Neurogenic hypertension.* London: Blackwell.

196 Dr Riamondo, V. C. and Forsham, P. H. (1958). Pharmacophysiologic principles in the use of corticoids and adrenocorticotrophin. *Metabolism,* 7: 5-24.

197 Dirken, J. M. (1966). Industrial shift works: Decrease in well being and specific effects. *Ergonomics,* 9: 115-124.

198 Dissmann, E. (1947). Klinishe Studien über den 24-Stunden-Rhythmus und seine Störungen bei Lungestuberkulösen. *Acta tuberc. scand.* 21: 204-240.

199 Dissmann, E. (1950). Lungenblutung und 24-Stunden-Rhythmus. *Acta. tuberc. scand.* 24: 277-292.

200 Dissmann, E. (1950). Zur Frage von Eigenrhythmus und Grundrhythmus in den Tagesschwankungen der Vitalkapazität. *Acta med. scand.* 137 :441-451.

201 Djavid, I. (1935). Über die Tagesschwankungen der Eosinophilenzahlen im Blut und die Beeinflussung der Eosinophilen durch Adrenalin. *Klin. Wschr.* 14: 930-931.

202 Doe, R. (1961). Cushing's Syndrome. In *Circadian Systems,* Report 39th Ross Conference on Pediatric Research, Fomon, S. J. (ed.). Columbus, Ohio: Ross Laboratories, 62-64.

203 Doe, R. P., Flink, E. B. and Flint, M. G. (1954). Correlation of diurnal variations in eosinophils and 17-hydroxycorticosteroids in plasma and urine. *J. clin. Endocrin.* 14: 774-775.

204 Doe, R. P., Flink, E. B. and Goodsell, M. G., (1956). Relationship of diurnal variation in 17-hydroxycorticosteroid levels in blood and urine to eosinophils and electrolyte excretion. *J. clin. Endocrin.* 16: 196-206.

205 Doe, R. P., Vennes, J. A. and Flink, E. B. (1960). Diurnal variation of 17-hydroxycorticosteroids, sodium, potassium, magnesium and creatinine in normal subjects and in cases of treated adrenal insufficiency and Cushing's Syndrome. *J. clin. Endocrin.* 20: 253-265.

206 Doig, R. J., Mummery, R. V., Wills, M. R. and Elkes, A. (1966). Plasma cortisol levels in depression. *Brit. J. Psychiat.* 112: 1263-1267.

207 Doll, R., Jones, F. Avery and Buckatzsch, M. M. (1951). Occupational factors in the aetiology of gastric and duodenal ulcers, with an estimate of their incidence in the general population. *Medical Research Council, Special Report Series,* No. 276, London.

208 Dollery, C. T., Emslie-Smith, D., and Milne, M. D. (1960). Clinical and pharmacological studies with guanethidine in the treatment of hypertension. *Lancet,* 2: 381-387.

209 Donaldson, I. M., Doig, A. and Knight, I. C. (1967). Nephroptosis with nocturnal polyuria. *Amer. J. Med.* 43: 289-293.

210 Donato, R. A. and Strumia, M. (1952). An exact method for the chamber count of eosinophils in capillary blood and its application to the study of the diurnal cycle. *Blood,* 7: 1020-1029.

211 Döring, G. K. and Riecke, H. (1952). Über tagesperiodische Schwankungen der Kapillarresistenz. *Klin. Wschr.* 30: 1098-1100.

212 Döring, G. K. and Schaefers, E. (1950). Über die Tagesrhythmik der Pupillenweite beim Menschen. *Pflüg. Arch. ges. Physiol.* 252: 537-541.

213 Dorman, P. J., Sullivan, W. J., Pitts, R. F., Lange, S. F. and MacLeod, M. B. (1954). The renal response to acute respiratory acidosis. *J. clin. Invest.* 33: 82-90.

214 Dossetor, J. B., Gorman, H. M. and Beck, J. C. (1963). The diurnal rhythm of urinary electrolyte excretion. I. Observations in normal subjects. *Metabolism*, 12: 1083-1099.

215 Drance, S. M. (1960). The significance of the diurnal tension variations in normal and glaucomatous eyes. *Arch. Ophthal.* 64: 494-501.

216 Dray, F., Reinberg, A. and Sebaoun, J. C. R. (1965). Rythme biologique de la testostérone libre du plasma chez l'homme adulte sain: existence d'une variation circadienne. *C.R. Acad. Sci., Paris*, 261: 573-576.

217 Dreyer, G., Bazett, H. C. and Pierce, H. F. (1920). Diurnal variation in the hemoglobin content of the blood. *Lancet*, 2: 588-591.

218 Dublin, W. B., Gregg, R. O. and Broders, A. C. (1940). Mitosis in specimens removed during day and night from carcinoma of large intestine. *Arch. Path.* 30: 893-895.

219 Dubois, E. L. and Alder, D. C. (1963). Single-daily dose oral administration of corticosteroid in rheumatic disorders. *Curr. Ther. Res.* 5: 43-56.

220 Duke-Elder, S. (1952). The phasic variations in the ocular tension in primary glaucoma. *Amer. J. Ophthal.* 35: 1-21.

221 Duke-Elder, S. (1955). Clinical aspects of the outflow of the aqueous humour. III. The phasic variations. In *Glaucoma.* A symposium. Oxford: Blackwell. 147-159.

222 Ehret, C. F. and Barlow, J. S. (1960). Toward a realistic model of a biological period-measuring mechanism. *Cold Spr. Harb. Symp. quant. Biol.* 25: 217-220.

223 Ehret, C. F. and Trucco, E. (1967). Molecular models for the circadian clock. I. The chronon concept. *J. theoret. Biol.* 15: 240-262.

224 Eik-Nes, K. and Clark, L. D. (1958). Diurnal variation of plasma hydroxycorticosteroids in subjects suffering from severe brain damage. *J. clin. Endocrin.* 18: 764-768.

225 Eik-Nes, K. B., Oertel, G. W., Nimer, R. and Tyler, F. H. (1959). Effect of human chorionic gonadotrophin on plasma concentrations of 17-hydroxycorticosteroids, dehydroepiandrosterone and androsterone in man. *J. clin. Endocrin.* 19: 1405-1410.

226 Ekman, H., Hakansson, B., McCarthy, J. D., Lehmann, J. and Sjögren, B. (1961). Plasma 17-hydroxycorticosteroids in Cushing's Syndrome. *J. clin. Endocrin.* 21: 684-694.

227 Elithorn, A., Bridges, P. K., Lobban, M. C. and Tredre, B. E. (1966). Observations on some diurnal rhythms in depressive illness. *Brit. med. J.* 2: 1620-1623.

228 Elliot, J. S., Sharp, R. F. and Lewis, L. (1959). Urinary pH. *J. Urol.* 81: 339-343.

229 Elliott, A. L. and Mills, J. N. (1969). Urinary potassium rhythms before and after transatlantic flight. *J. Physiol.* 200: 122P.

230 Elmadjian, F. and Pincus, G. (1946). A study of the diurnal variation in

circulating lymphocytes in normal and psychotic subjects. *J. clin. Endocrin.* 6: 287-294.

231 ELMSLIE, R. G., WHITE, T. T. and MAGEE, D. F. (1964). Observations on pancreatic function in eight patients with controlled pancreatic fistulas: including a review of the literature. *Ann. Surg.* 160: 937-949.

232 ELWOOD, P. C. (1962). Diurnal haemoglobin variations in normal male subjects. *Clin. Sci.* 23: 379-382.

233 ENDRES, G. (1922). Über Gesetzmässigkeiten in der Beziehung zwischen der wahren Harnreaktion und der alveolaren CO_2—Spannung. *Biochem. Z.* 132: 220-241.

234 ENDRES, G. (1923). Atmungsregulation und Blutreaktion im Schlaf. *Biochem. Z.* 142: 53-67.

235 ENGEL, R., HALBERG, F. and GULLY, R. J. (1952). The diurnal rhythm in EEG discharge and in circulating eosinophils in certain types of epilepsy. *Electroenceph. clin. Neurophysiol.* 4: 115-116.

236 ENRIGHT, J. T. (1965). Accurate geophysical rhythms and frequency analysis. In *Circadian Clocks*. Amsterdam: North Holland. 31-42.

237 ERICSON, L. A. (1958). Twenty-four hourly variations in the inflow of the aqueous humour. *Acta ophthal., Kbh.* Supp. 50: 1-95.

238 ERNEST, I. (1966). Steroid excretion and plasma cortisol in 41 cases of Cushing's Syndrome. *Acta. endocr., Copenhagen*, 51: 511-525.

239 ERTEL, R. J., UNGAR, F. and HALBERG, F. (1963). Circadian rhythm in susceptibility of mice to toxic doses of SU-4885. *Med. Proc.* 22: 211.

240 FAIMAN, C. and MOORHOUSE, J. A. (1967). Diurnal variation in the levels of glucose and related substances in healthy and diabetic subjects during starvation. *Clin. Sci.* 32: 111-126.

241 FAIMAN, C. and RYAN R. J. (1967). Diurnal cycle in serum concentrations of follicle-stimulating hormone in men. *Nature, Lond.* 215: 857.

242 FEARNLEY, G. R., BALMFORTH, G. and FEARNLEY, E. (1957). Evidence of a diurnal fibrinolytic rhythm; with a simple method of measuring natural fibrinolysis. *Clin. Sci.* 16: 645-650.

243 FEIGIN, R. D., KLAINER, A. S. and BEISEL, W. R. (1967). Circadian periodicity of blood-amino acids in adult men. *Nature, Lond.* 215: 512-514.

244 FEIGIN, R. D., KLAINER, A. S. and BEISEL, W. R. (1968). Factors affecting circadian periodicity of blood amino acids in man. *Metabolism*, 17: 764-775.

245 FINK, B. R. (1961). Influence of cerebral activity in wakefulness on regulation of breathing. *J. appl. Physiol.* 16: 15-20.

246 FINKENSTAEDT, J. T., DINGMAN, J. F., JENKINS, D., LAIDLAW, J. C. and MERRILL, J. P. (1954). The effect of intravenous hydrocortisone and corticosterone on the diurnal rhythm in renal function and electrolyte equilibria in normal and Addisonian patients. *J. clin. Invest.* 33: 933.

247 FINLAYSON, D. C., DAGHER, F. J. and VANDAM, L. D. (1964). Diurnal variation in blood volume of man. *J. Surg. Res.* 4: 286-288.

248 FISHER, L. B. (1968). The diurnal mitotic rhythm in the human epidermis. *Brit. J. Derm.* 80: 75-80.

249 FISHER, B. and FISHER, E. R. (1951). Observations on the eosinophil

count in man: a proposed test of adrenal cortical function. *Amer. J. med. Sci.* 221: 121-132.

250 FISKE, C. H. (1921). Inorganic phosphate and acid excretion in the post-absorptive period. *J. Biol. Chem.* 49: 171-181.

251 FLINK, E. B. and DOE, R. P. (1959). Effect of sudden time displacement by air travel on synchronization of adrenal function. *Proc. Soc. exp. Biol., N.Y.* 100: 498-501.

252 FLINK, E. B. and HALBERG, F. (1952). Clinical studies on eosinophil rhythm. *J. clin. Endocrin.* 12: 922.

253 FOMON, S. J. (1961). *Circadian Systems*: Report of the thirty-ninth Ross Conference on Pediatric Research. Columbus, Ohio: Ross Laboratories.

254 FONTÈS, G. and YOVANOVITCH, A. (1923). Influence du sommeil sur l'élimination des principaux composés azotés. *C.R. Soc. Biol.* 88: 456-458.

255 FORSGREN, E. (1930). Vierundzwanzig-Stunden-Variationen der Gallensekretion. *Scand. Arch. Physiol.* 59: 217-225.

256 FORSGREN, E. (1947). To what degree does gastric acidity show a tendency to rise or fall during the day? *Acta med. scand.* 128: 281-288.

257 FORSHAM, P. H., DI RIAMONDO, V., ISLAND, D., RINFRET, A. P. and ORR, R. H. (1955). Dynamics of adrenal function in man. In *The Human Adrenal Cortex*. Wolstenholme, G. E. W., Cameron, M. P. and Etherington, J. (Eds.). *Ciba Found. Colloq. Endocrin.* 8: 279-308.

258 FORT, A. (1968). Circadian alterations in alertness. *J. Physiol.* 197: 82-83P.

259 FORT, A. (1969). The effects of rapid change in time zone on circadian variation in psychological functions. *J. Physiol.* 200: 124P.

260 FOX, R. H., GOLDSMITH, R., HAMPTON, E. F. G. and WILKINSON, R. T. (1963). The effects of a raised body temperature on the performance of mental tasks. *J. Physiol.* 167: 22-23P.

261 FRAISSE, P., SIFFRE, M., OLERON, G. and ZUILI, N. (1968). Le rythme veille-sommeil et l'estimation du temps. In *Cycles Biologiques et Psychiatrie*, de Ajuriaguerra, J. (ed). Symposium Bel-Air III. Geneva: Georg. 257-265.

262 FRAJOLA, W. J. (1961). Circadian aspects of enzyme activity. In *Circadian Systems*, Report 39th Ross Conference on Pediatric Research, Fomon, S. J. (ed.), Columbus, Ohio: Ross Laboratories, 74-75.

263 FRANK, G., HARNER, R., MATTHEWS, J., JOHNSON, E., and HALBERG, F. (1961). Circadian periodicity and the human electroencephalogram. *Electroceph. clin. Neurophysiol.* 13: 822.

264 FRANKS, R. C. (1967). Diurnal variation of plasma 17-hydroxycorticosteroids in children. *J. clin. Endocrin.* 27: 75-78.

265 FRAZIER, T. W., RUMMEL, J. A. and LIPSCOMB, H. S. (1968). Circadian variability in vigilance performance. *Aerospace Med.* 39: 383-395.

266 FREINKEL, N., MAGER, M. and VINNICK, L. (1968). Cyclicity in the interrelationships between plasma insulin and glucose during starvation in normal young men. *J. Lab. clin. Med.* 71: 171-178.

267 FREY, S. (1929). Der Tod des Menschen in seinen Beziehungen zu den Tages und Jahreszeiten. *Dtsch. Z. Chir.* 218: 336-369.

268 FRIEDMAN, A. H. and WALKER, C. A. (1968). Circadian rhythms in rat mid-brain and caudate nucleus biogenic amine levels. *J. Physiol.* 197: 77-85.

269 FRIEDRICH, L., GARAI, T. and FALUCHI, E. A. (1958). Quantitative

Untersuchungen der Tagesschwankungen der Urobilinogenausscheidung. *Z. ges. inn. Med.* 13: 900-903.
270 Fröberg, J. Personal communication.
271 Ganong, W. F., Shepherd, M. D., Wall, J. R., van Brunt, E. E. and Clegg, M. T. (1963). Penetration of light into the brain of mammals. *Endocrinology*, 72: 962-963.
272 Garcia-Sainz, M. and Halberg, F. (1966). Mitotic rhythms in human cancer, re-evaluated by electronic computer programs—evidence for temporal pathology. *J. natn. Cancer Inst.* 37: 279-292.
273 Garrod, O. and Burston, R. A. (1952). The diuretic response to ingested water in Addison's disease and panhypopituitarism and the effect of cortisone thereon. *Clin. Sci.* 11: 113-128.
274 Gates, A. I. (1910). Diurnal variations in memory and association. *University of California Publications in Psychology*. 1.
275 Gates, A. I. (1916). Variations in efficiency during the day. *University of California Publications in Psychology*. 2.
276 Gauquelin, M. (1967). Note sur le rythme journalier du début des douleurs de l'accouchement. *Gynec. et Obstet.* 66: 229-236.
277 Gazenko, O. G. (1964). Medical investigations of spaceships Vostok and Voskhod. *3rd Int. Sym. on Bioastronautics and the exploration of Space. San Antonio, Texas.*
278 Gemzell, C. A. (1953). Blood levels of 17-hydroxycorticosteroids in normal pregnancy. *J. clin. Endocrin.* 13: 898-902.
279 Gerritzen, F. (1936). Liver diuresis. *Acta. med. scand.* 89: 101-123.
280 Gerritzen, F. (1962). The diurnal rhythm in water, chloride, sodium and potassium excretion during a rapid displacement from East to West and vice versa. *Aerospace Med.* 33: 697-701.
281 Gerritzen, F. (1966). Influence of light on human circadian rhythms. *Aerospace Med.* 37: 66-70.
282 Ghata, J. and Reinberg, A. (1954). Variations nycthémérale, saisonniere et geographique de l'élimination urinare du potassium et de l'eau chez l'homme adulte sain. *C.R. Acad, Sci., Paris*, 239: 1680-1682.
283 Gibbons, J. L. (1964). Cortisol secretion rate in depressive illness. *Arch. gen. Psychiat.* 10: 572-575.
284 Gibbons, J. L. and McHugh, P. R. (1962). Plasma cortisol in depressive illness. *J. Psychiat. Res.* 1: 162-171.
285 Gibson, R. B. (1905). The effects of transposition of the daily routine on the rhythm of temperature variation. *Amer. J. med. Sci.* 129: 1048-1059.
286 Gifford, S. (1960). Sleep, time and the early ego. *J. Amer. Psychoanal. Ass.* 8: 5-42.
287 Gill, J. R., Jr., Melmon, K. L., Gillespie, L., Jr., and Bartter, F. C. (1965). Bradykinin and renal function in normal man: effects of adrenergic blockade. *Amer. J. Physiol.* 209: 844-848.
288 Givens, J. R., Ney, R. L., Nicholson, W. E., Graber, A. L. and Liddle, G. W. (1964). Absence of a normal diurnal variation of plasma ACTH in Cushing's Disease. *Clin. Res.* 12: 267.
289 Gjessing, R. (1936). Beiträge zur Kenntnis der Pathophysiologie der katotonen Erregung, mit kritischem Beginn und Abschluss. *Arch. Psychiat.* 104: 355-416.

290 GLICKS, S. M. and GOLDSMITH, S. (1967). The physiology of growth hormone secretion. In *Growth Hormone*. Int. Cong. Series. No. 158, 84-88. Excerpa. Medica Foundation, Amsterdam.

291 GOETZL, F. R. and STONE, F. (1947). Diurnal variations in acuity of olfaction and food intake. *Gastroenterology*, 9: 444-453.

292 GOLDECK, H. (1948). Der 24-Stunden-Rhythmus der Erythropoese. *Ärztl-Forsch.* 2: 22-27.

293 GOLDECK, H., HERRNRING, G. and RICHTER, U. (1950). Die 24-Stunden-Periodik der Thrombozyten. *Dtsch. med. Wschr.* 75: 702-703.

294 GOLDECK, H. and SIEGEL, P. (1948). Die 24-Stunden-Periodik der Blutreticulocyten unter vegetativen Pharmaka. *Ärztl. Forsch.* 2: 245-248.

295 GOLDMAN, R. (1951). Studies in diurnal variation of water and electrolyte excretion: nocturnal diuresis of water and sodium in congestive cardiac failure and cirrhosis of the liver. *J. clin. Invest.* 30: 1191-1199.

296 GOLDMAN, R. and LUCHSINGER, E. (1956). Relationship between diurnal variations in urinary volume and the excretion of antidiuretic substance. *J. clin. Endocrin.* 16: 28-34.

297 GOLDSMITH, R. S., SIEMSEN, A. W., MASON, A. D., Jr., and FORLAND, M. (1965). Primary role of plasma hydrocortisone concentration in the regulation of the normal forenoon pattern of urinary phosphate excretion. *J. clin. Endocrin.* 25: 1649-1659.

298 GOODWIN, B. C. (1963). *Temporal Organisation in Cells.* London: Academic Press.

299 GOODWIN, B. C. (1966). An entrainment model for timed enzyme synthesis in bacteria. *Nature, Lond.* 209: 479-481.

300 GOODWIN, J. C., JENNER, F. A., LOBBAN, M. C. and SHERIDAN, M. (1964). Renal rhythms in a patient with a 48-hour cycle of psychosis during a period of life on an abnormal time routine. *J. Physiol.* 176: 16P.

301 GOODWIN, J. C., JENNER, F. A. and SLATER, S. E. (1968). The diurnal pattern of excretion of antidiuretic hormone. *J. Physiol.* 196: 112-113P.

302 GORDON, R. D., KUCHEL, O., ISLAND, D. P. and LIDDLE, G. W. (1966). Role of sympathetic nervous system in mediating renal and adrenocortical secretory responses to upright posture. *J. clin. Invest.* 45: 1016.

303 GORDON, R. D., SPINKS, J., DULMANIS, A., HUDSON, B., HALBERG, F. and BARTTER, F. C. (1968). Amplitude and phase relations of several circadian rhythms in human plasma and urine: demonstration of rhythm for tetrahydrocortisol and tetrahydrocorticosterone. *Clin. Sci.* 35: 307-324.

304 GORDON, R. D., WOLFE, L. K., ISLAND, D. P. and LIDDLE, G. W. (1966). A diurnal rhythm in plasma renin activity in man. *J. clin. Invest.* 45: 1587-1592.

305 GORDON, T. (1964). Glucose tolerance of adults. *National Center for Health Statistics, Series* 11. *No.* 2. *U.S. Public Health Service Publication.*

306 GOWENLOCK, A. H., MILLS, J. N. and THOMAS, S. (1959). Acute postural changes in aldosterone and electrolyte excretion in man. *J. Physiol.* 146: 133-141.

307 GRABER, A. L., GIVENS, J. R., NICHOLSON, W. E., ISLAND, D. P. and LIDDLE, G. W. (1965). Persistence of diurnal rhythmicity in plasma ACTH concentrations in cortisol-deficient patients. *J. clin. Endocrin.* 25: 804-807.

308 Graber, A. L., Ney, R. L., Nicholson, W. E., Island, D. P. and Liddle, G. W. (1965). Natural history of pituitary-adrenal recovery following long-term suppression with corticosteroids. *J. clin. Endocrin.* 25; 11-16.

309 Grant, S. D., Pavlatos, F. C. and Forsham, P. H. (1965). Effects of estrogen therapy on cortisol metabolism. *J. clin. Endocrin.* 25: 1057-1066.

310 Grether, W. F. (1965). Human performance for military and civilian operations in space. *Ann. N.Y. Acad. Sci.* 134: 398-412.

311 Griffiths, G. M. and Fox, J. T. (1938). Rhythm in epilepsy. *Lancet,* 2: 409-416.

312 Grossman, L. I. and Brickman, B. M. (1937). Comparison of diurnal and nocturnal pH values of saliva. *J. dent. Res.* 16: 179-182.

313 Guberan, E., Williams, M. K., Walford, J. and Smith, M. M. (1969). Diurnal variation of FEV in shiftworkers. *Brit. J. industr. Med.* 26: 121-125.

314 Gunn, H. E., Jr., Unger, A. L., Hume, D. M. and Schilling, J. A. (1960). Human renal transplantation: an investigation of the functional status of the denervated kidney after successful homotransplantation in identical twins. *J. Lab. clin. Med.* 56: 1-13.

315 Günther, R., Knapp, E. and Halberg, F. (1968). Circadiane Rhythmometrie mittels elektronischer Rechner zur Beurteilung von Kurwirkungen. In *Kurverlaufs und Kurerfolgsbeurteilung*; Symposium II. Teichmann, W. (ed.), Bad Worishofen: Sanitas-Verlag, 106-111.

316 Günther, R., Knapp, E. and Halberg, F. (1969). Referenznormen der Rhythmometrie: Circadiane Acrophasen von zwanzig Körperfunktionen. *Sonderband der Zeitschrift fur Bader—und Klimaheilkunde.* Schattauer Verlag: Stuttgart. In Press.

317 Gutteridge, J. M. (1968). Thin-layer chromatographic techniques for the investigation of abnormal urinary catecholamine metabolite patterns. *Clin. Chim. Acta.* 21: 211-216.

318 Hague, E. B. (1964). Photo-neuroendocrine effect in circadian systems with particular reference to the eye. *Ann. N.Y. Acad. Sci.* 117: 1-645.

319 Hajjar, G. C., Whissen, N. C. and Moser, K. M. (1961). Diurnal variation in plasma euglobulin activity and fibrinogen. *Angiology,* 12: 160-164.

320 Halberg, F. (1953). Some physiological and clinical aspects of 24-hour periodicity. *J. Lancet,* 73: 20-32.

321 Halberg, F. (1959). Physiologic 24-hour periodicity: General and procedural considerations with reference to the adrenal cycle. *Z. Vitam—Horm.—u. Fermentforsch.* 10: 225-296.

322 Halberg, F. (1960). Temporal coordination of physiologic function. *Cold Spr. Harb. Symp. quant. Biol.* 25: 289-310.

323 Halberg, F. (1960). The 24-hour scale: a time dimension of adaptive functional organization. *Persp. Biol. Med.* 3: 491-527.

324 Halberg, F. (1963). Periodicity analysis. A potential tool for biometeorologists. *Intern. J. Biometeor.* 7: 167-191.

325 Halberg, F. (1963). Circadian (about twenty-four-hour) rhythms in experimental medicine. *Proc. R. Soc. Med.* 56: 253-260.

326 Halberg, F. (1964). Organisms as circadian systems; temporal analysis of their physiologic and pathologic responses, including injury and death. In *Walter Reed Army Institute of Research Symposium, Medical Aspects of Stress in Military Climate.* 1-36.

327 HALBERG, F. (1965). Some aspects of biologic data analysis; longitudinal and transverse profiles of rhythms. In *Circadian Clocks*. Aschoff, J. (ed.), Amsterdam: North Holland, 13-22.

328 HALBERG, F. (1969). Chronobiology. *Ann. Rev. Physiol.* 31: 675-725.

329 HALBERG, F. Personal communication.

330 HALBERG, F., ANDERSON, J. A., ERTEL, R. and BERENDES, H. (1967). Circadian rhythm in serum 5-hydroxytryptamine of healthy men and male patients with mental retardation. *Int. J. Neuropsychiat.* 3: 379-386.

331 HALBERG, F., BITTNER, J. J., GULLY, R. J., ALBRECHT, P. G. and BRACHNE, E. L. (1955). 24-hour periodicity and audiogenic convulsions in I mice of various ages. *Proc. Soc. exp. Biol.* N.Y. 88: 169-173.

332 HALBERG, F., COHEN, S. L. and FLINK, E. B. (1951). Two new tools for the diagnosis of adrenal cortical dysfunction. *J. Lab. clin. Med.* 38: 817.

333 HALBERG, F., ENGEL, R., HALBERG, E. and GULLY, R. J. (1952). Diurnal variations in amount of electroencephalographic paroxysmal discharge and diurnal eosinophil rhythm of epileptics on days with clinical seizures. *Fed. Proc.* 11: 62.

334 HALBERG, F., ENGEL, R., TRELOAR, A. E. and GULLY, R. J. (1953). Endogenous eosinopenia in institutionalized patients with mental deficiency. *A.M.A. Arch. Neurol. Psychiat*, 69: 462-469.

335 HALBERG, F., ENGEL, M., HAMBURGER, C. and HILLMAN, D. (1965). Spectral resolution of low frequency, small amplitude rhythms in excreted 17-ketosteroids; probably androgen induced circaseptan desynchronization. *Acta. endocr., Copenhagen Supp.* 103: 5-54.

336 HALBERG, F., FLINK, E. B. and VISSCHER, M. B. (1951). Alterations in diurnal rhythm in circulating eosinophil levels in adrenal insufficiency. *Amer. J. Physiol.* 167: 791.

337 HALBERG, F., FRENCH, L. and GULLY, R. J. (1958). 24-hour rhythms in rectal temperature and blood eosinophils after hemidecortication in human subjects. *J. appl. Physiol.* 12: 381-384.

338 HALBERG, F., HALBERG, E., BARNUM, C. P. and BITTNER, J. J. (1959). Physiologic 24-hour periodicity in human beings and mice, the lighting regimen, and daily routine. In *Photoperiodism and related phenomena in plants and animals*. Washington, D.C. *Am. Assoc. Advan. Sci. Publ.* 55: 803-878.

339 HALBERG, F., HALBERG, E. and GULLY, R. J. (1953). Effects of modifications of the daily routine in healthy subjects and in patients with convulsive disorder. *Epilepsia*. 3rd series, 2: 150.

340 HALBERG, F. and HAMBURGER, C. (1964). 17-ketosteroid and volume of human urine. Weekly and other changes with low frequency. *Minn. Med.* 47: 916-925.

341 HALBERG, F. and HOWARD, R. B. (1958). 24-hour periodicity and experimental medicine; examples and interpretations. *Postgrad. Med.* 24: 349-358.

342 HALBERG, F., JACOBSON, E., WADSWORTH, G. and BITTNER, J. J. (1958). Audiogenic abnormality spectra, 24-hour periodicity and lighting. *Science*, 128: 657-658.

343 HALBERG, F., JOHNSON, E. A., BROWN, B. W. and BITTNER, J. J. (1960).

Susceptibility rhythm to E. coli endotoxin and bioassay. *Proc. Soc. exp. Biol. N.Y.* 103: 142-144.

344 HALBERG, F. and PANOFSKY, H. (1961). Thermo-Variance spectra: method and clinical illustrations. *Exp. Med. Surg.* 19: 284-309.

345 HALBERG, F. and PANOFSKY, H. (1961). Thermo-Variance spectra: Simplified computational example and other methodology. *Exp. Med. Surg.* 19: 323-338.

346 HALBERG, F., SIFFRE, M., ENGELI, M., HILLMAN, D. and REINBERG, A. (1965). Étude en libre-cours des rythmes circadiens du pouls, de l'alternance veille-sommeil et de l'estimation du temps pendant les deux mois de séjour souterrain d'un homme adulte jeune. *C.R. Acad. Sci., Paris,* 260: 1259-1262.

347 HALBERG, F., TONG, Y. L. and JOHNSON, E. A. (1967). Circadian system phase—an aspect of temporal morphology; procedures and illustrative examples. In *The Cellular Aspects of Biorhythms,* Von Mayersbach, H. (ed.). Symposium on Biorhythms. Berlin: Springer-Verlag. 20-48.

348 HALBERG, F. and ULSTROM, R. A. (1952). Morning changes in number of circulating eosinophils in infants. *Proc. Soc. exp. Biol.,* N.Y. 80: 747-748.

349 HALBERG, F., VISSCHER, M. B., FLINK, E. B., BERGE, K. and BOCK, F. (1951). Diurnal rhythmic changes in blood eosinophil levels in health and in certain diseases. *J. Lancet,* 71: 312-319.

350 HAMBY, G. W., PRANGE, A. J., Jr., LIPTON, M. A. and COCHRANE, C. M. (1963). Physiologic variations of a tyramine-like substance in human saliva. *J. Nerv. Ment. Dis.* 137: 487-493.

351 HAMILTON, L. D., GUBLER, C. J., CARTWRIGHT, G. E. and WINTROBE, M. M. (1950). Diurnal variation in the plasma iron level of man. *Proc. Soc. exp. Biol., N.Y.* 75: 65-68.

352 HAMPP, H. (1961). Die Tagesrhythmischen Schwankungen der Stimmung und des Antriebes beim gesunden Mensch. *Arch. Psychiat. Nervenkr.* 201: 355-377.

353 HARKER, J. E. (1958). *The Physiology of Diurnal Rhythms.* Cambridge: Cambridge Univ. Press.

354 HARKER, J. E. (1958). Diurnal rhythms in the animal kingdom. *Biol. Rev.* 33: 1-52.

355 HARRIS, J. D. and MYERS, C. K. (1954). Experiments on fluctuation of auditory acuity. *J. gen. Psychol.* 50: 87-109.

356 HART, P. D'A. and VERNEY, E. B. (1934). Observations on the rate of water loss by man at rest. *Clin. Sci.* 1: 367-396.

357 HARTER, J. G., REDDY, W. J. and THORN, G. W. (1963). Studies on an intermittent corticosteroid dosage regimen. *New Engl. J. Med.* 269: 591-596.

358 HASTINGS, J. W. (1959). Unicellular clocks. *Ann. Rev. Microbiol.* 13: 297-312.

359 HASTINGS, J. W. (1968). Biochemical mechanisms involved in biological rhythms and cycles. In *Cycles Biologiques et Psychiatrie,* de Ajuriaguerra, J. (ed.). Symposium Bel-Air III. Geneva: Georg. 127-140.

360 HASTINGS, J. W. and KENYAN, A. (1965). Molecular aspects of circadian systems. In *Circadian Clocks,* Aschoff, J. (ed.), Amsterdam: North Holland.

361 HAUS, E. and HALBERG, F. (1966). Circadian phase diagrams of oral

temperature and urinary functions in a healthy man studied longitudinally. *Acta. endocr., Copenhagen* 51: 215-223.

362 HAUTY, G. T. and ADAMS, T. (1966). Phase shifts of the human circadian system and performance deficit during the periods of transition: I. East-West flight. *Aerospace Med.* 37: 668-674.

363 HAUTY, G. T. and ADAMS, T. (1966). Phase shifts of the human circadian system and performance deficit during the periods of transition: II. West-East flight. *Aerospace Med.* 37: 1027-1033.

364 HAUTY, G. T. and ADAMS, T. (1966). Phase shifts of the human circadian system and performance deficit during the periods of transition. III. North-South flight. *Aerospace Med.* 37: 1257-1262.

365 HAWKING, F. (1955). Periodicity of microfilariae of Loa loa. *Trans. Roy. soc. trop. Med. Hyg.* 49: 132-142.

366 HAWKING, F. (1960). Periodicity of microfilariae. *Indian J. Malar.* 14: 567-574.

367 HAWKING, F. (1963). Circadian rhythms in filariasis. *Proc. R. Soc. Med.* 56: 260.

368 HAWKING, F. (1964). The periodicity of microfilariae. VIII. Further observations on Wucheria bancrofti. *Trans. Roy. soc. trop. Med. Hyg.* 58: 212-227.

369 HAYWOOD, B. J. and STARKWEATHER, W. H. (1968). Diurnal variation of erythrocyte glucose-6-phosphate dehydrogenase and 6-phosphogluconate dehydrogenase. *Amer. J. clin. Path.* 49: 275-278.

370 HEATON, F. W. and HODGKINSON, A. (1963). External factors affecting diurnal variation in electrolyte excretion with particular reference to calcium and magnesium. *Clin. chim. Acta.* 8: 246-254.

371 HEILMEYER, L. and PLÖTNER, K. (1937). *Das Serumeisen und die Eisenmangelkrankheit.* Jena: Fischer.

372 HELLBRÜGGE, T. (1960). The development of circadian rhythms in infants. *Cold Spr. Harb. Symp. quant. Biol.* 25: 311-323.

373 HELLBRÜGGE, T. (1967). Chronophysiologie des Kindes. *Verh. dtsch. Ges. inn. Med.* 73: 895-921.

374 HELLBRÜGGE, T. (1968). Ontógènese des rythmes circadiaires chez l'enfant. In *Cycles Biologiques et Psychiatrie*, de Ajuriaguerra, J. (ed.). Symposium Bel-Air III. Geneva: Georg. 159-183.

375 HELLBRÜGGE, T., EHRENGUT-LANGE, J. and RUTENFRANZ, J. (1956). Über die Entwicklung von Tagesperiodischen Veränderungen der Pulsfrequenz im Kindesalter. *Z. Kinderhk.* 78: 703-722.

376 HELLBRÜGGE, T., EHRENGUT-LANGE, J., RUTENFRANZ, J. and STEHR, K. (1964). Circadian periodicity of physiological functions in different stages of infancy and childhood. *Ann. N.Y. Acad. Sci.* 117: 361-373.

377 HELLEBRAND, F. A., TEPPER, R. H., GRANT, H. and CATHERWOOD, R. (1936). Nocturnal and diurnal variations in the acidity of the spontaneous secretion of gastric juice. *Amer. J. dig. Dis.* 3: 477-481.

378 HEMMELER, G. (1943). Le fer sérique dans les ictères parenchymateux et par rétention. *Schweiz. med. Wschr.* 73: 1056.

379 HEMMELER, G. (1944). Nouvelles recherches sur le métabolisme du fer; les oscillations du fer sérique dans la journée. *Helv. med. Acta.* 11: 201-207.

380 HENDRICKS, S. B. (1963). Metabolic control of timing. *Science*, 141: 21-27.

381 HERBERT, P. J. and DE VRIES, J. A. (1949). The administration of adrenocorticotrophic hormone to normal human subjects. The effect of the leucocytes in the blood and on circulating antibody level. *Endocrinology*, 44: 259-273.

382 HILDEBRANDT, G. (1954). Bäderkuren und biologische Rhythmen. *Dtsch. med. Wschr.* 79: 1404-1405.

383 HILDEBRANDT, G. (1955). Untersuchungen über die rhythmische Functions-ordnung von Puls und Atem. *Acta. med. Scand. Supp.* 307: 175-184.

384 HILDEBRANDT, G. (1962). Reaktive Perioden und Spontanrhythmik. *Atti. VII. Conf. Inter. Soc. Stud. Ritm. Biol.* Turin: Panminerva Med. 75-82.

385 HILDEBRANDT, G. (1967). Rhythmus und Regulation unter besonderer Berücksichtigung der Blutdruckregulation. *Z. ges. Inn. Med. 22: Suppl:* 206-213.

386 HILDEBRANDT, G. and ENGELBERTZ, P. (1953). Bedeutung der Tagesrhythmik für die physikalische Therapie. *Arch. phys. Ther. Lpz.* 5: 160-170.

387 HILLS, A. G., FORSHAM, P. H. and FINCH, C. A. (1948). Changes in circulating leucocytes induced by the administration of pituitary adrenocorticotrophic hormone (ACTH) in man. *Blood*, 3: 755-768.

388 HIMMEL, G. K., MARTHALER, T. M., RATEITSCHAK, K. H. and MÜHLEMANN, H. R. (1957). Experimental changes of diurnal periodicity in the physical properties of periodontal structures. *Helv. odont. Acta*, 1: 16-18.

389 HIRSCH, H. M. (1961). Interrelationship of the genetic, the biochemical, and the circadian components. In *Circadian Systems*, Report 39th Ross Conference on Pediatric Research, Fomon, S. J. (ed.), Columbus, Ohio: Ross Laboratories. 71-73.

390 HOAGLAND, H. (1933). The physiological control of judgements of duration: evidence for a chemical clock. *J. gen. Psychol.* 9: 267-287.

391 HOAGLAND, H., PINCUS, G., FRED, E., ROMANOFF, L., FREEMAN, H., HOPE, J., BALLAN, J., BERKELEY, A. and CARLO, J. (1953). Study of adrenocortical physiology in normal and schizophrenic men. *Arch. Neurol. Psychiat., Chicago*, 69: 470-485.

392 HOBBS, G. E., GODDARD, E. S. and STEVENSON, J. A. F. (1954). The diurnal cycle in blood eosinophils and body temperature. *Canad. med. Assoc. J.* 70: 533-536.

393 HOHENEGGER, M. (1966). Day-night excretion rhythms of adrenal cortical steroids in kidney disorders and nocturia. *Endokrinologie*, 50: 260-266.

394 HOHENEGGER, M. and KAISER, E. (1967). Tagesrhythmische Ausscheidung der alkalischen Harnphosphatase. *Klin. Wschr.* 45: 1252-1253.

395 HOKFELT, B. and LUFT, R. (1959). The effect of suprasellar tumours on the regulation of adrenocortical function. *Acta. endocr., Copenhagen*, 32: 177-186.

396 HOLLINGWORTH, H. L. (1914). Variations in efficiency during the working day. *Psychol. Rev.* 21: 473-491.

397 HOLLISTER, L. E. (1963). Successive trials of induced alimentary lipemia. *Amer. J. clin. Nutr.* 13: 214-218.

398 HOLLISTER, L. E. and WRIGHT, A. (1965). Diurnal variation of serum lipids. *J. Atheroscler. Res.* 5: 445-450.

399 HOLLWICH, F. (1964). The influence of light via the eyes on animals and man. *Ann. N.Y. Acad. Sci.* 117:105-131.

400 HOLLWICH, F. and DIECKHUES, B. (1968). Eosinopeniereaktion und sehvermögen. *Klin. MBL Augenheilk.* 152: 11-16.

401 HOLMGREN, H., MÖLLERSTRÖM, J. and SWENSSON, A. (1953). Verhandl. III Konf. Internat. Gesellsch. Biol. Rhythmusforschung, Hamburg, 1949. *Acta. Med. scand. Suppl.* 278: 1-146.

402 HOOPER, J. H. (1866). The variations of temperature in health. *Med. Times Hosp. Gaz.* 2: 483-484.

403 HOPKINS, H. (1933). Time of appearance of epileptic seizures in relation to age, duration and type of syndrome. *J. Nerv. Ment. Dis.* 77: 153-162.

404 HOSKINS, R. G. (1946). *The Biology of Schizophrenia.* The Thomas William Memorial Lectures of the New York Academy of Medicine. New York: Norton.

405 HOWELL, W. H. (1897). A contribution to the physiology of sleep based upon plethysmographic experiments. *J. exp. Med.* 2: 313-346.

406 HOWITT, J. S., BALKWILL, J. S., WHITESIDE, T. C. D. and WHITTINGHAM, P. D. G. (1966). A preliminary study of flight deck work loads in civil air transport aircraft. *U.K. Ministry of Defence (Air Force Dept.) FPRC.* 1240.

407 HØYER, K. (1944). Physiologic variations in the iron content of human blood serum. I. The variations from week to week, from day to day and through twenty-four hours. *Act. med. scand.* 119: 562-576.

408 HØYER, K. (1944). Physiologic variations in the iron content of human blood serum. II. Further studies of the intra diem variations. *Acta. med. scand.* 119: 577-585.

409 HUGUENIN, S. (1899). Die Späterfolge der Glaukombehandlung bei 76 Privatpatienten von Prof. Dr. Haab, Zurich. *Deutschmanns Beitr. prakt. Augenheilk.* 32: 1-97.

410 HULLIN, R. P., BAILEY, A. D., MCDONALD, R., DEANSFIELD, D. A. and MILNE, H. B. (1967). Variations in 11-hydroxycorticosteroids in depression and manic-depressive psychosis. *Brit. J. Psychiat.* 113: 593-600.

411 HUNTER, W. M., FRIEND, J. A. R. and STRONG, J. A. (1966). The diurnal pattern of plasma growth hormone concentration in adults. *J. Endocrin.* 34: 139-146.

412 HUNTER, W. M. and GREENWOOD, F. C. (1962). A radio-immunoelectrophoretic assay for human growth hormone. *Biochem. J.* 85: 39-40P.

413 HUNTER, W. M. and GREENWOOD, F. C. (1964). A radio-immunoelectrophoretic assay for human growth hormone. *Biochem. J.* 91: 43-56.

414 HUNTER, W. M. and RIGAL, W. M. (1966). The diurnal pattern of plasma growth hormone concentration in children and adolescents. *J. Endocrin.* 34: 147-153.

415 HURWICZ, (1962). Basic mathematical and statistical considerations in the study of rhythms and near rhythms. *Ann. N.Y. Acad. Sci.* 98: 851-857.

416 IBERALL, A. S. and CARDON, S. Z. (1964). Control in biological systems—a physical review. *Ann. N.Y. Acad. Sci.* 177: 445-518.

417 ILETT, K. F. and LOCKETT, M. F. (1968). A renally active substance from heart muscle and from blood. *J. Physiol.* 196: 101-109.

418 IMRIE, M. J., MILLS, J. N. and WILLIAMSON, K. S. (1963). The renal action of small doses of cortisol at night. *J. Endocrin.* 27: 289-292.

419 IMRIE, M. J., MILLS, J. N. and WILLIAMSON, K. S. (1963). Circadian variations in renal and adrenal function: Are they connected? Hormones and the Kidney. *Mem. Soc. Endocrin.* 13: 3-13, London and New York: Academic Press.

420 JACEY, M. J. and SCHAEFER, K. E. (1968). Circadian cycles of lactic dehydrogenase in urine and blood plasma: response to high pressure. *U.S. Navy Submarine Center, Groton, Conn. Rep. No.* 534.

421 JACKLET, J. W. (1969). Circadian rhythm of optic nerve impulses recorded in darkness from isolated eye of Aplysia. *Science,* 164: 562-563.

422 JAEGER, H. (1881). Über die Korperwärme des gesunden Menschen. *Dtsch. Arch. Klin. Med.* 29: 516-536.

423 JANSEN, G., RUTENFRANZ, J. and SINGER, R. (1966). Über eine circadiane Rhythmik sensumotorischer Leistungen. *Int. Z. angew. Physiol.* 22: 65-83.

424 JAPHA, A. (1900). Die Leukocyten beim gesunden und kranken Säugling. *Jahrb. f. Kinderhlk.* 52: 242-270.

425 JAROS, M. and KOTULAN, J. (1964). Mental efficiency in school-children in shift lessons. *Activ. Nerv. Su. (Praha).* 6: 96.

426 JENNER, F. A., GOODWIN, M., SHERIDAN, I. J., TAUBER, J. and LOBBAN, M. C. (1968). The effect of an altered time regime on biological rhythms in a 48 hour periodic psychosis. *Brit. J. Psychiat.* 114: 215-224.

427 JENNY, E. (1933). Tagesperiodische Einflüsse auf Geburt und Tod. *Schweiz. med. Wschr.* 63: 15-17.

428 JOFFE, A., MAINZER, F. and SCHERER, E. (1929). Über das Zeitliche Verhalten der Ausscheidung organischer Säuren im Harn. *Z. Klin. Med.* 111: 464-471.

429 JOHNSTON, C. I. and DAVIS, J. L. (1966). Evidence from cross circulation studies for a humoral mechanism in the natriuresis of saline loading. *Proc. Soc. exp. Biol., N.Y.* 121: 1058-1062.

430 JOHNSTON, R. L. and WASHEIM, H. (1924). Studies in gastric secretion. II. Gastric secretion in sleep. *Amer. J. Physiol.* 70: 247-253.

431 JONES, H. BENCE. (1845). Contributions to the chemistry of the urine. On the variations in the alkaline and earthy phosphates in the healthy state, and on the alkalescence of the urine from fixed alkalies. *Phil. Trans.* 135: 335-349.

432 JONES, E. LLOYD, (1889). On the variations in the specific gravity of the blood in health. *J. Physiol.* 8: 1-14.

433 JONES, E. LLOYD, (1891). Further observations on the specific gravity of the blood in health and disease. *J. Physiol.* 12: 299-346.

434 JONES, N. F., BARRACLOUGH, M. A. and MILLS, I. H. (1963). Mechanism of increased sodium excretion during water loading with 2.5 per cent dextrose and vasopressin. *Clin. Sci.* 25: 449-457.

435 JORES, A. (1933). Die Urineinschränkung in der Nacht. *Dtsch. Arch. Klin. Med.* 175: 224-253.

436 JORES, A. (1934). Die 24-Stundenperioden des Menschen. *Med. Klin.* 30: 1-468.

437 JORES, A. (1935). Physiologie und Pathologie der 24-Stunden-Rhythmik des Menschen. *Ergebn. inn. Med. Kinderheilk.* 48: 574-629.

438 JORES, A. (1938). Zur Rhythmusforschung. *Dtsch. med. Wschr.* 64: 737-748.

439 JORES, A. (1938). Endokrines und Vegetatives System in ihrer Bedeutung für die Tagesperiodik. *Dtsch. med. Wschr.* 64: 989-993.

440 JORES, A. (1940). Verhandl. II Konf. Internat. Gessellsch. Biol. Rhythmusforschung. *Acta. med. scand. Suppl.* 108: 23-243.

441 JORES, A. and FREES, J. (1937). Die Tagesschwankungen der Schmerzempfindung. *Dtsch. med. Wschr.* 63: 962-963.

442 JOSEPHSON, B. and LARSON, H. (1934). Über der Periodizität der Gallensekretion bei einem Patienten mit Gallenfistel. *Scand. Arch. Physiol.* 69: 227-236.

443 JUNDELL, J. (1904). Über die nykthemeralen Temperaturschwankungen im 1. Lebensjahr des Menschen. *Jahrb. f. Kinderheilk.* 59: 521-619.

444 JÜRGENSEN, T. (1867). Zur Lehre von der Behandlung fieberhafter Krankheiten mittelst des kalten Wassers. *Dtsch. Arch. Klin. Med.* 3: 165-222.

445 JUSATZ, J. H. and ECKHARDT, E. (1934). Die häufigste Todesstunde. *Münch. Med. Wschr.* 81: 709-710.

446 KAINE, H. D., SELTZER, H. S. and CONN, J. W. (1955). Mechanism of diurnal eosinophil rhythm in man. *J. Lab. clin. Med.* 45: 247-252.

447 KAISER, I. H. (1961). Circadian aspects of human birth. In *Circadian Systems*, Report 39th Ross Conference on Pediatric Research, Fomon, S. J. (ed.), Columbus, Ohio: Ross Laboratories. 33-34.

448 KAISER, I. H. and HALBERG, F. (1962). Circadian periodic aspects of birth. *Ann. N.Y. Acad. Sci.* 98: 1056-1058.

449 KAISER, F. and MAURATH, J. (1949). Kreislaufdynamische 24-Stunden-Rhythmik beim Menschen. *Klin. Wschr.* 27: 659-662.

450 KALMUS, H. and WIGGLESWORTH, L. A. (1960). Shock excited systems as models for biological rhythms. *Cold Spr. Harb. Symp. quant. Biol.* 25: 211-216.

451 KANEKO, M., ZECHMAN, F. W. and SMITH, R. E. (1968). Circadian variation in human peripheral blood flow levels and exercise responses. *J. appl. Physiol.* 25: 109-114.

452 KÄRKI, N. T. (1956). The urinary excretion of noradrenaline and adrenaline in different age groups, its diurnal variation and the effect of muscular work on it. *Acta Physiol. scand. 39 Supp.* 132: 1-96.

453 KATAVISTO, M. (1964). The diurnal variations of ocular tension in glaucoma. *Acta. Ophthal. Supp.* 78: 1-131.

454 KATZ, F. H. (1964). Adrenal function during bed rest. *Aerospace Med.* 35: 849-851.

455 KAYE, G. (1929). Studies in the reaction of urine. *Aust. J. exper. Biol. med. Sci.* 6: 187-214.

456 KELLGREN, J. H. and JANUS, O. (1951). The eosinopenic response to cortisone and ACTH in normal subjects. *Brit. med. J.* 2: 1183-1187.

457 KERKHOFF, J. F. and PETERS, J. H. (1968). A reproducible estimation of the urobilin concentration in urine by means of a modified Schlesinger test. *Clin. chim. Acta.* 21: 133-137.

458 KILINSKII, E. L. and EGART, F. M. (1964). Coronary circulation in patients with diabetes mellitus (diurnal variations in ECG patterns). *Fed. Proc. Transl. Suppl.* 23: 301-303.

459 King, P. D. (1956). Increased frequency of births in the morning hours. *Science*, 123: 985-986.

460 King, P. D. (1960). Distortion of the birth frequency curve. *Amer. J. Obstet. Gynec.* 79: 399-400.

461 Kirschner, M. A., Lipsett, M. B. and Collins, D. R. (1965). Plasma ketosteroids and testosterone in man: A study of the pituitary-testicular axis. *J. clin. Invest.* 44: 657-665.

462 Kleber, R. S., Lhamon, W. T. and Goldstone, S. (1963). Hyperthermia, hyperthyroidism, and time judgement. *J. Comp. Physiol. Psychol.* 56: 362-365.

463 Klein, K. E., Brüner, H., Ruff, S. and Godesberg, B. (1966). Untersuchungen zur Belastung des Bordpersonals auf Fernflügen mit Düsenmaschinen. *Zeit. für Flugwissenschaften.* 14: 109-121.

464 Klein, K. E., Wegmann, H. M. and Brüner, H. (1968). Circadian rhythm in indices of human performance, physical fitness and stress resistance. *Aerospace Med.* 39: 512-518.

465 Kleitman, N. (1923). Studies on the physiology of sleep. 1. The effects of prolonged sleeplessness on man. *Amer. J. Physiol.* 66: 67-92.

466 Kleitman, N. (1949). Biological rhythms and cycles. *Physiol. Rev.* 29: 1-30.

467 Kleitman, N. (1949). The sleep-wakefulness cycle in submarine personnel. In *National Research Council. Human Factors in Undersea Warfare*. Baltimore: Waverly Press. 329-341.

468 Kleitman, N. (1952). Sleep. *Sci. Amer.* 187: 34-38.

469 Kleitman, N. (1961). Development of circadian rhythm in the infant. In *Circadian Systems*, Report 39th Ross Conference on Pediatric Research, Fomon, S. J. (ed.), Columbus, Ohio: Ross Laboratories. 35-37.

470 Kleitman, N. (1963). *Sleep and Wakefulness*, 2nd Ed. Chicago and London: University of Chicago Press.

471 Kleitman, N. (1967). Phylogenetic, ontogenetic and environmental determinants in the evolution of sleep-wakefulness cycles. *Res. Publ. Ass. nerv. ment. Dis.* 45: 30-38.

472 Kleitman, N. and Doktorsky, A. (1933). The effect of the position of the body and of sleep on rectal temperature in man. *Amer. J. Physiol.* 104: 340-343.

473 Kleitman, N. and Engelmann, T. G. (1953). Sleep characteristics of infants. *J. appl. Physiol.* 6: 269-282.

474 Kleitman, N. and Jackson, D. P. (1950). Body temperature and performance under different routines. *J. appl. Physiol.* 3: 309-328.

475 Kleitman, N. and Kleitman, E. (1953). Effect of non-twenty-four-hour routines of living on oral temperature and heart rate. *J. appl. Physiol.* 6: 283-291.

476 Kleitman, N. and Ramsaroop, A. (1948). Periodicity in body temperature and heart rate. *Endocrinology*, 43: 1-20.

477 Kleitman, N., Titelbaum, S. and Feiveson, P. (1938). The effect of body temperature on reaction time. *Amer. J. Physiol.* 121: 495-501.

478 Kleitman, N., Titelbaum, S. and Hoffman, H. (1937). The establishment of the diurnal temperature cycle. *Amer. J. Physiol.* 119: 48-54.

479 Knapp, C. B. (1909). The hour of birth. *Bull. Lying-in-Hosp. N.Y.* 6: 69-74.

480 KNAPP, M. S., KEANE, P. M., and WRIGHT, J. C. (1967). Circadian rhythm of plasma 11-hydroxycorticosteroids in depressive illness, congestive heart failure, and Cushing's syndrome. *Brit. med. J.* 2: 27-30.

481 KOBAYASHI, T., LOBOTSKY, J. and LLOYD, C. W. (1966). Plasma testosterone and urinary 17-ketosteroids in Japanese and Occidentals. *J. clin. Endocrin.* 26: 610-614.

482 KOJIMA, A. and NIIYAMA, Y. (1965). Diurnal variations of 17-ketogenic steroid and catecholamine excretion in adolescent and middle-aged shift workers with special reference to adaptability to night work. *Indust. Health.* 3: 9-19.

483 KRANZFELD, B. (1925). Zur Frage über die physiologischen Tagesschwankungen der Thrombocytenzahl. *Pflüg. Arch. ges. Physiol.* 210: 583-585.

484 KRIEGER, (1869). Über die Entstehung von Entzündlichen und fieberhafter Krankheiten. *Z. Biol.* 5: 476-535.

485 KRIEGER, D. T. (1961). Diurnal pattern of plasma 17-hydroxycorticosteroids in pretectal and temporal lobe disease. *J. clin. Endocrin.* 21: 695-698.

486 KRIEGER, D. T. and KRIEGER, H. P. (1966). Circadian variation of the plasma 17-hydroxycorticosteroids in central nervous system disease. *J. clin. Endocrin.* 26: 929-940.

487 KROETZ, C. (1926). Über einige stoffliche Erscheinungen bei verlängertem Schlaftentzug. *Z. ges. exp. Med.* 52: 770-778.

488 KROETZ, C. (1940). Ein biologischer 24-Stunden-Rhythmus des Blutkreislaufs bei Gesundheit und bei Herzschwäche, zugleich ein Beitrag zur tageszeitlichen Häufung einiger akuter Kreislaufstörungen. *Münch. med. Wschr.* 87: I. 284-288.

489 KROETZ, C. (1940). Ein biologischer 24-Stunden-Rhythmus des Blutkrieslaufs bei Gesundheit und bei Herzschwäche, zugleich ein Beitrag zur tageszeitlichen Häufung einiger akuter Kreislaufstörungen. *Münch. med. Wschr.* 87: II. 314-317.

490 KUO, P. T. and CARSON, J. C. (1959). Dietary fats and the diurnal serum triglyceride levels in man. *J. clin. Invest.* 38: 1384-1393.

491 LAFONTAINE, E., GHATA, J., LAVERNHE, J., COURILLON, J., BELLANGER, G. and LAPLANE, R. (1967). Rhythmes biologiques et décalages horaires (I). Étude expérimentale au cours de vols commerciaux long-courriers. *Le Concours Med.* 89; 3731-3746.

492 LAFONTAINE, E., GHATA, J., LAVERNHE, J., COURILLON, J., BELLANGER, G. and LAPLANE, R. (1967). Rythmes biologiques et décalages horaires. (II). Étude expérimentale au cours de vols commerciaux longs courriers. *Le Concours Med.* 89: 3963-3976.

493 LÄHR, H. (1889). Versuche über den Einfluss des Schlafes auf den Stoffwechsel. *Allg. Ztschr. f. Psychiat.* 46: 286-317.

494 LAIDLAW, J. C., JENKINS, D., REDDY, W. J. and JAKOBSON, T. (1954). The diurnal variation in adrenocortical secretion. *J. clin. Invest.* 33: 950.

495 LAIDLAW, J. C., REDDY, W. J., JENKINS, D., HAYDAR, N. A., RENOLD, A. E. and THORN, G. W. (1955). Advances in the diagnosis of altered states of adrenocortical function. *New Engl. J. Med.* 253: 747-753.

496 LAIRD, D. A. (1925). Relative performance of college students as conditiohed by time of day and day of week. *J. exp. Psychol.* 8: 50-63.

497 Lamb, E. J., Dignam, W. J., Pion, R. J. and Simmer, H. H. (1964). Plasma androgens in women. *Acta. endocr. Copenhagen.* 45: 243-253.

498 Lambert, A. E. and Hoet, J. J. (1965). Diurnal pattern of plasma insulin concentration in the human. *Diabetologia*, 2: 69-71.

499 Landau, J. and Feldman, S. (1954). Diminished endogenous morning eosinopenia in blind subjects. *Acta. endocr. Copenhagen.* 15: 53-60.

500 Langdon-Down, M. and Brain, W. R. (1929). Time of day in relation to convulsions in epilepsy. *Lancet*, 1: 1029-1032.

501 Langley, D. and Swanljung, H. (1951). Ocular tension in glaucoma simplex. *Brit. J. Ophthal.* 35: 445-458.

502 Lavernhe, J., Lafontaine, E. and Laplane, R. (1965). Subjective effects of time shifts (an inquiry among flight personnel of Air France). *Rev. Méd. Aéronautique.* 4: 30-36.

503 Layne, D. S., Meyer, C. J., Vaishwanar, P. S. and Pincus, G. (1962). The secretion and metabolism of cortisol and aldosterone in normal and in steroid-treated women. *J. clin. Endocrin.* 22: 107-118.

504 Leake, C. D., Kohl, M. and Stebbins, G. (1927). Diurnal variations in the blood specific gravity and erythrocyte count in healthy human adults. *Amer. J. Physiol.* 81: 493.

505 Leathes, J. B. (1919). Renal efficiency tests in nephritis and the reaction of the urine. *Brit. med. J.* 2: 165.

506 Levi, L. (1966). Physical and mental stress reactions during experimental conditions simulating combat. *Sartr. Försvarsmedicin.* 2: 3-8.

507 Levin, E., Kirsner, J. B., Palmer, W. L. and Butler, C. (1948). The variability and periodicity of the nocturnal gastric secretion in normal individuals. *Gastroenterology*, 10: 939-951.

508 Levine, H., Ramshaw, W. A. and Halberg, F. (1967). Least squares spectral analyses on core temperature and blood pressure of a comatose girl. *Physiologist*, 10: 230.

509 Levy, F. M. and Conge, G. (1953). Action de la lumière sur l'éosinophilie sanguine chez l'homme. *C. R. Soc. Biol.* 147: 586-589.

510 Lewinsohn, H. C., Capel, L. H. and Smart, J. (1960). Changes in forced expiratory volumes throughout the day. *Brit. Med. J.* 1: 462-464.

511 Lewis, A. A. G. (1953). The control of the renal excretion of water. *Ann. R. Coll. Surg. Engl.* 13: 36-54.

512 Lewis, B. M., McElroy, W. T., Hayford-Welsing, E. J. and Samberg, L. C. (1960). The effects of body position, ganglionic blockade, and norepinephrine on the pulmonary capillary bed. *J. clin. Invest.* 39: 1345-1352.

513 Lewis, P. R. and Lobban, M. C. (1956). Patterns of electrolyte excretion in human subjects during a prolonged period of life on a 22-hour day. *J. Physiol.* 133: 670-680.

514 Lewis, P. R. and Lobban, M. C. (1957). The effects of prolonged periods of life on abnormal time routines upon excretory rhythms in human subjects. *Quart. J. exp. Physiol*, 42: 356-371.

515 Lewis, P. R. and Lobban, M. C. (1957). Dissociation of diurnal rhythms in human subjects living on abnormal time routines. *Quart. J. exp. Physiol.* 42: 371-386.

516 Lewis, P. R. and Lobban, M. C. (1958). The effects of exercise on diurnal excretory rhythms in man. *J. Physiol.* 143: 9P.

517 Lewis, P. R., Lobban, M. C. and Shaw, T. I. (1956). Patterns of urine flow in human subjects during a prolonged period of life on a 22-hour day. *J. Physiol.* 133: 659-669.

518 Lhamon, W. T. (1968). Psychopathology and conceptions of time. In *Cycles Biologiques et Psychiatrie*, de Ajuriaguerra, J. (ed.), Symposium Bel-Air III, Geneva: Georg. 281-289.

519 Lichardus, B. and Pearce, J. W. (1966). Evidence for a humoral natriuretic factor released by blood volume expansion. *Nature, Lond.* 209: 407-409.

520 Lichter, I. and Sirett, N. E. (1968). Plasma cortisol levels in lung cancer. *Brit. med. J.* 2: 154-156.

521 Liddle, G. W. (1966). Analysis of circadian rhythms in human adrenocortical secretory activity. *Arch. intern. Med.* 117: 739-743.

522 Liddle, G. W., Island, D. and Meador, C. K. (1962). Normal and abnormal regulation of corticotrophin secretion in man. *Recent Progr. Hormone Res.* 18: 125-166.

523 Lindan, O., Baker, W. R. Jr., Greenway, R. M., King, P. H., Piazza, J. M. and Reswick, J. B. (1965). Metabolic rhythms of the quadriplegic patient. 1. Effect of rhythmic and random feeding and body turning schedule on the hourly excretion pattern of urinary metabolites. *Arch. phys. Med.* 46: 79-88.

524 Lindhard, J. (1911). Conditions governing the body temperature (leading article in Brit. med. J. referring to Danmark Expeditionen Grünlands, Nordöstkyst. 1906-1908. Vol. IV, No. 1). *Brit. med. J.* 1: 829-830.

525 Lindheimer, M. D., Lalone, R. C. and Levinsky, N. G. (1967). Evidence that acute increase in glomerular filtration has little effect on sodium excretion in dog unless extracellular volume is expanded. *J. clin. Invest.* 46: 256-266.

526 Lindsay, A. E., Migeon, C. J., Nugent, C. A. and Brown, H. (1956). The diagnostic value of plasma and urinary 17-hydroxycorticosteroid determinations in Cushing's syndrome. *Amer. J. Med.* 20: 15-22.

527 Lines, J. G. and Ranger, I. (1969). Diurnal rhythms of pancreatic function. *J. Physiol.* 200: 57-58P.

528 Linnér, E. (1957). The effect of prednisolone on aqueous humour dynamics. *Acta. Soc. Med. upsalien.* 62: 186-192.

529 Linnér, E. (1958). 24-hourly variations in the outflow of the aqueous humour. *Acta. ophthal.* 36: 381-385.

530 Linnér, E. (1959). The rate of aqueous flow and the adrenals. *Trans. ophth. Soc. U.K.* 79: 27-32.

531 Linquette, M., Fossati, P., Racadot, A., Hubschman, B., and Decoulx, M. (1968). Les variations circadiennes de la cortisolemie dans la maladie de Cushing et les états frontières. *Ann. Endocrin., Paris*, 29: 69-76.

532 Lipsett, M. B., Wilson, H., Kirschner, M. S., Korenman, S. G., Fishman, L. M., Sarfaty, G. A. and Bardin, C. W. (1966). Studies on Leydig cell physiology and pathology: secretion and metabolism of testosterone. *Recent Progr. Hormone Res.* 22: 245-281.

533 LOBBAN, M. C. (1960). The entrainment of circadian rhythms in man. *Cold Spr. Harb. Symp. quant. Biol.* 25: 325-332.

534 LOBBAN, M. C. (1963). Human renal diurnal rhythms in an Arctic mining community. *J. Physiol.* 165: 75-76P.

535 LOBBAN, M. C. (1965). Dissociation in human rhythmic function. In *Circadian Clocks*. Aschoff, J. (ed.). Amsterdam: North Holland. 219-227.

536 LOBBAN, M. C. (1967). Daily rhythms of renal excretion in arctic-dwelling Indians and Eskimos. *Quart. J. exp. Physiol.* 52: 401-410.

537 LOBBAN, M. C. and TREDRE, B. E. (1964). Renal diurnal rhythms in human subjects during bed-rest and limited activity. *J. Physiol.* 171: 26-27P.

538 LOBBAN, M. C. and TREDRE, B. E. (1966). Daily rhythms of renal excretion in human subjects with irregular hours of work. *J. Physiol.* 186: 139-140P.

539 LOBBAN, M. C. and TREDRE, B. E. (1967). Diurnal rhythms of renal excretion and of body temperature in aged subjects. *J. Physiol.* 188: 48-49P.

540 LOBBAN, M. C. and TREDRE, B. E. (1967). Perception of light and the maintenance of human renal diurnal rhythms. *J. Physiol.* 189: 32-33P.

541 LOBBAN, M. C., TREDRE, B., ELITHORN, A. and BRIDGES, P. (1963). Diurnal rhythms of electrolyte excretion in depressive illness. *Nature, Lond.* 199: 667-669.

542 LOCKETT, M. F. (1967). Effects of salt loading and haemodilution on the responses of perfused cat kidneys to angiotensin. *J. Physiol.* 193: 639-647.

543 LOCKETT, M. F. (1967). Hormonal actions of the heart and of lungs on the isolated kidney. *J. Physiol.* 193: 661-669.

544 LOCKETT, M. F. & ROBERTS, C. N. (1963). Hormonal factors affecting sodium excretion in the rat. *J. Physiol.* 167: 581-590.

545 LOCKETT, M. F. and ROBERTS, C. N. (1963). Some actions of growth hormone on the perfused cat kidney. *J. Physiol.* 169: 879-888.

546 LOCKNER, D. (1966). The diurnal variation of plasma iron turnover and erythropoiesis in healthy subjects and cancer patients. *Brit. J. Haemat.* 12: 646-656.

547 LONGSON, D. and MILLS, J. N. (1953). The failure of the kidney to respond to respiratory acidosis. *J. Physiol.* 122: 81-92.

548 LONGSON, D., MILLS. J. N., THOMAS, S. and YATES, P. A. (1956). Handling of phosphate by the human kidney at high plasma concentrations. *J. Physiol.* 131: 555-571.

549 LORAINE, J. A. and BELL, E. T. (1966). *Hormone Assays and their Clinical Application.* Edinburgh and London: Livingstone.

550 LOUTIT, J. F. (1965). Diurnal variation in urinary excretion of calcium and strontium. *Proc. Roy. Soc.* 162: 458-472.

551 LOVELAND, N. T. and WILLIAMS, H. L. (1963). Adding, sleep loss, and body temperature. *Percept. and Mot. Skills.* 16: 923-929.

552 LUETSCHER, J. A., DOWDY, A. J., ARNSTEIN, A. R., LUCAS, C. P. and MURRAY, C. L. (1964). Idiopathic oedema and increased aldosterone excretion. In *Aldosterone.* A symposium by the Council for International organizations of Medical Sciences. Baulieu, E. E. and Robel, P. (ed.). Oxford: Blackwell.

553 LUETSCHER, J. A. and LIEBERMAN, A. H. (1958). Aldosterone. *Arch. intern. Med.* 102: 314-330.

554 LUNEDEI, A., CAGNONI, M., FANTINI, F., TARQUINI, B., MORACE, G., MARELLO, M., PANERAI, H., SCARPELLI, P. T. and TOCCAFONDI, R. (1967). Sindromi di Encefaliche (Problemi in Discussioni). *Edizione Vione L. Pozzi, Roma.*

555 LURIE, A. O., REID, D. E. and VILLEE, C. A. (1966). The role of the fetus and placenta in maintenance of plasma progesterone. *Amer. J. Obstet. Gynec.* 96: 670-675.

556 MACKWORTH, N. H. (1950). Researches in the measurement of human performance. *Medical Research Council Special Report Series No.* 268. H.M.S.O. London.

557 MAGNUSSEN, G. (1944). *Studies on the Respiration during Sleep, a Contribution to the Physiology of the Sleep Function.* London: Lewis.

558 MAHAFFEY, J. H. and HAYNES, B. W. (1953). Observations on pancreatic juice in a case of external pancreatic fistula in man. *Amer. Surg.* 19: 174-181.

559 MÁLEK, J. (1952). The manifestation of biological rhythms in delivery. *Gynaecologia,* 133; 365-372.

560 MÁLEK, J. (1954). Der Einfluss des Lichtes und der Dunkelheit auf den klinischen Geburtsbeginn. *Gynaecologia,* 138: 401-405.

561 MÁLEK, J. (1962). The daily rhythm of the onset of labour, of excretion of pregnandiol, of white blood cell count, of blood pressure, pulse rate and temperature in pregnancy and the daily rhythm of the beginning of menstruation. *Atti VII Conf. Intern. Soc. Stud. Ritm. Biol.* Turin: Panminerva Med. 97-103.

562 MÁLEK, J. (1963). The daily rhythm of the human lactation. *8th International Conference of the Society for Biological Rhythm.* Hamburg, 1963.

563 MÁLEK, J., BUDINSKY, J. and BUDINSKA, M. (1950). Analyse du rythme journalière du début clinique de l'accouchement. *Rev. franc. Gynéc. Obstet.* 45: 222.

564 MÁLEK, J., GLEICH, J. and MALÝ, V. (1962). Characteristics of the daily rhythm of menstruation and labor. *Ann. N.Y. Acad. Sci.* 98: 1042-1055.

565 MÁLEK, J., ŠUK, K., BŘEŠŤÁK, M. (1962). Daily rhythm of leukocytes, blood pressure, pulse rate, and temperature during pregnancy. *Ann. N.Y. Acad. Sci.* 98: 1018-1041.

566 MANCHESTER, R. C. (1933). The diurnal rhythm in water and mineral exchange. *J. clin. Invest.* 11: 995-1008.

567 MANN, A. and LEHMANN, H. (1952). The eosinophil level in psychiatric conditions. *Canad. med. Ass. J.* 66: 52-58.

568 MANN, R. D. (1967). Effect of age, sex, and diurnal variation on the human fibrinolytic system. *J. clin. Path.* 20: 223-226.

569 MARSHALL, J. (1959). Alterations in the diurnal excretion of electrolytes during intermittent positive-pressure respiration. *Brit. med. J.* 2. 85-88.

570 MARTEL, P. J., SHARP, G. W. G., SLORACH, S. A. and VIPOND, H. J. (1962). A study of the roles of adrenocortical steroids and glomerular filtration rate in the mechanism of the diurnal rhythm of water and electrolyte excretion. *J. Endocrin.* 24: 159-169.

571 MARTIN, M. M. and HELLMAN, D. E. (1964). Temporal variation in

SU 4885 responsiveness in man: evidence in support of circadian variation in ACTH secretion. *J. clin. Endocrin.* 24: 253-260.

572 Martin, M. M. and Martin, A. L. A. (1968). Simultaneous fluorometric determination of cortisol and corticosterone in human plasma. *J. clin. Endocrin.* 28: 137-145.

573 Martin, M. M., Mintz, D. H. and Tamagaki, H. (1963). Effect of altered thyroid function upon steroid circadian rhythms in man. *J. clin. Endocrin.* 23: 242-247.

574 Maruhama, Y., Goto, Y. and Yamagata, S. (1967). Diabetic treatment and the diurnal plasma triglyceride. *Metabolism,* 16: 985-995.

575 Maslenikow, A. (1904). Über Tagesschwankungen des intraokulären Druckes bei Glaukom. *Z. für Aug.* 11: 564.

576 Masoni, A. (1960). 24-hour rhythms of cardiovascular functions in normal and hypertensive subjects. *Atti. VII Conf. Intern. Soc. Stud. Ritm. Biol.* Turin: Panminerva Med. 110.

577 Massarrat, S. (1964). Tagesschwankungen der Enzymaktivitäten der Glutamat-Oxalacetat, Glutamat-Pyruvat Transaminasen und Lactatdehydrogenase im Serum bei Leberkranken und Gesunden. *Klin. Wschr.* 42: 91-94.

578 Mauer, A. M. (1965). Diurnal variation of proliferative activity in the human bone marrow. *Blood,* 26: 1-7.

579 Meddis, R. (1968). Human circadian rhythms and the 48-hour day. *Nature, Lond.* 218: 964-965.

580 Mellette, H., Hutt, B. K., Askovitz, S. I. and Horvath, S. M. (1951). Diurnal variations in body temperature. *J. appl. Physiol.* 3: 665-675.

581 Menzel, W. (1941). Der 24-Stunden-Rythmus des menschlichen Blutkreislaufes. *Ergebn. inn. Med. Kinderheilk.* 61: 1-53.

582 Menzel, W. (1950). Zur Physiologie und Pathologie des Nacht und Schichtarbeiters. *Arbeitsphysiologie,* 14: 304-318.

583 Menzel, W. (1953). Wellenlänge, Phasenlage und Amplitude in der Nierenrhythmik des Menschen. *Acta. med. scand. Suppl.* 278: 95-98.

584 Menzel, W. (1955). Klinische Ziele der Rhythmusforschung. *Acta. med. scand. Suppl.* 307: 107-116.

585 Menzel, W. (1962). Periodicity in urinary excretion in healthy and nephropathic persons. *Ann. N.Y. Acad. Sci.* 98: 1007-1017.

586 Menzel, W. (1962). *Menschliche Tag-Nacht-Rhythmik und Schichtarbeit.* Basel and Stuttgart: Benno Schwabe & Co.

587 Menzel, W. (1966). Rhythmen im Krankheitsverlauf. *Med. Klin.* 61: 201-203.

588 Menzel, W., Blume, J. and Lua, E. (1953). Untersuchungen zur Nierenrhythmik. *Z. ges. Exp. Med.* 120: 396-410.

589 Menzel, W., Möllerström, J. and Petrén, T. (1955). Verhandl. IV Konf. Intern. Gesellsch. Biol. Rhythmusforschung. Basle, 1953. *Acta med. scand. Suppl.* 307: 1-199.

590 Menzel, W., Timm, R. and Herrnring, G. (1949). Über den diagnostischen Wert der Tag-Nacht-Schwankungen des erhöhten Blutdruckes. *Verh. dtsch. Ges. Kreisl. Forsch.* 15: 256-260.

591 Mercer, D. M. A. (1960). Analytical methods for the study of periodic

phenomena obscured by random fluctuations. *Cold Spr. Harbor Symp. quant. Biol.* 25: 73-86.

592 MERTZ, VON D. P. and ISELE, W. (1964). Tagesperiodische Anderungen der Dynamik des endogenen Jodstoffwechsels. *Med. Klin.* 59: 1536-1539.

593 METZ, B. and ANDLAUER, P. (1949). Le rythme nycthéméral de la température chez l'homme. *C.R. Soc. Biol., Paris.* 143: 1234-1236.

594 METZ, B. and SCHWARTZ, J. (1949). Études des variations de la température rectale, du debit urinaire et de l'excretion urinaire des 17-cetosteroides chez l'homme au cours du nycthémere. *C.R. Soc. Biol., Paris.* 143: 1237-1239.

595 MEYER, R. J. (1953). Relative insensitivity of the hypothalmic-pituitary-adrenal system to activation by epinephrine. *J. clin. Endocrin.* 13: 123-125.

596 MIGEON, C. J., BERTRAND, J. and WALL, P. E. (1957). Physiological disposition of 4-C^{14}-cortisol during late pregnancy. *J. clin. Invest.* 36: 1350-1362.

597 MIGEON, C. J., KELLER, A. R., LAWRENCE, B. and SHEPARD, T. H. (1957). Dehydroepiandrosterone and androsterone levels in human plasma. Effect of age and sex; day to day and diurnal variations. *J. clin. Endocrin.* 17: 1051-1062.

598 MIGEON, C. J., TYLER, F. H., MAHONEY, J. P., FLORENTIN, A. A., CASTLE, H., BLISS, E. L. and SAMUELS, L. T. (1956). The diurnal variation of plasma levels and urinary excretion of 17-hydroxycorticosteroids in normal subjects, night workers, and blind subjects. *J. clin. Endocrin.* 16: 622-633.

599 MIGEON, C., WALLACE, W. M. and METCOFF, J. (1961). Concluding remarks and implications in pediatrics. In *Circadian Systems*, Report 39th Ross Conference on Pediatric Research, Fomon, S. J. (ed.), Columbus Ohio: Ross Laboratories. 92-93.

600 MILLS, I. H. (1965). Effect of renal arteriolar tone in sodium excretion. *J. Endocrin.* 32 (supp): 4-5.

601 MILLS, I. H., SCHEDL, H. P., CHEN, P. S. Jr. and BARTTER, F. C. (1960). The effect of estrogen administration on the metabolism and protein binding of hydrocortisone. *J. clin. Endocrin.* 20: 515-528.

602 MILLS, J. N. (1946). Hyperpnoea induced by forced breathing. *J. Physiol.* 105: 95-116.

603 MILLS, J. N. (1951). Diurnal rhythm in urine flow. *J. Physiol.* 113: 528-536.

604 MILLS, J. N. (1953). Changes in alveolar carbon dioxide tension by night and during sleep. *J. Physiol.* 122: 66-80.

605 MILLS, J. N. (1955). The acute response to potassium ingestion. *J. Physiol.* 128: 47P.

606 MILLS, J. N. (1963). Diurnal variations in renal function. In *Recent Advances in Physiology*, Creese, R. (ed.). London: Churchill. 252-329.

607 MILLS, J. N. (1964). Circadian rhythms during and after three months in solitude underground. *J. Physiol.* 174: 217-231.

608 MILLS, J. N. (1966). Human circadian rhythms. *Physiol. Rev.* 46: 128-171.

609 MILLS, J. N. (1966). Sleeping habits during four months in solitude. *J. Physiol.* 189: 30-31P.

610 MILLS, J. N. (1967). Temperature and potassium excretion in a class experiment in circadian rhythmicity. *J. Physiol.* 194: 19P.

611 Mills, J. N. (1967). Keeping in step—away from it all. *New Scientist*, 33: 350-351.

612 Mills, J. N. (1968). Circadian rhythms. In *A Companion to Medical Studies*, Vol. 1. Passmore, R. and Robson, J. S. (eds.). Oxford: Blackwell.

613 Mills, J. N. and Stanbury, S. W. (1952). Persistent 24-hour renal excretory rhythm on a 12-hour cycle of activity. *J. Physiol.* 117: 22-37.

614 Mills, J. N. and Stanbury, S. W. (1954). A reciprocal relationship between K^+ and H^+ excretion in the diurnal excretory rhythm in man. *Clin. Sci.* 13: 177-186.

615 Mills, J. N. and Stanbury, S. W. (1955). Rhythmic diurnal variations in the behaviour of the human renal tubule. *Acta. med. scand. Suppl.* 307: 95-96.

616 Mills, J. N. and Thomas, S. (1957). The acute effect of adrenal hormones and carbohydrate metabolism upon plasma phosphate and potassium concentrations in man. *J. Endocrin.* 16: 164-179.

617 Mills, J. N. and Thomas, S. (1957). Diurnal excretory rhythms in a subject changing from night to day work. *J. Physiol.* 137: 65-66P.

618 Mills, J. N. and Thomas, S. (1958). The acute effects of cortisone and cortisol upon renal function in man. *J. Endocrin.* 17: 41-53.

619 Mills, J. N. and Thomas, S. (1959). The influence of adrenal corticoids on phosphate and glucose exchange in muscle and liver in man. *J. Physiol.* 148: 227-239.

620 Mills, J. N., Thomas, S. and Williamson, K. S. (1961). The effects of intravenous aldosterone and hydrocortisone on the urinary electrolytes of the recumbent human subject. *J. Physiol.* 156: 415-423.

621 Mills, J. N., Thomas, S. and Williamson, K. S. (1962). The extent of the adrenal influence upon renal electrolyte excretion in the healthy man on a normal diet, and its contribution to the renal changes on standing. *J. Endocrin.* 23: 365-373.

622 Mills, J. N., Thomas, S. and Yates, P. A. (1954). Reappearance of renal excretory rhythm after forced disruption. *J. Physiol.* 125: 466-474.

623 Min, H. K., Jones, J. E. and Flink, E. B. (1966). Circadian variations in renal excretion of magnesium, calcium, phosphorus, sodium, and potassium during frequent feeding and fasting. *Fed. Proc.* 25: 917-921.

624 Ministry of Labour, (Great Britain). (1965). *Shift working.* Gazette 73: 148-155. London: H.M.S.O.

625 Ministry of Munitions, (Great Britain). (1917). *Health of Munition Workers Committee.* Interim report. Industrial efficiency and fatigue. [Cd. 8511). London: H.M.S.O.

626 Ministry of Munitions (Great Britain). (1918). *Health of Munition Workers Committee.* Industrial Health and Efficiency. Final Report. (Cd. 9065). London: H.M.S.O.

627 Mintz, D. H., Hellman, D. E. and Canary, J. J. (1968). Effect of altered thyroid function on calcium and phosphorus circadian rhythms. *J. clin. Endocrin.* 28: 399-411.

628 Mishell, D. R., Jr., Wide, L. and Gemzell, C. A. (1963). Immunologic determination of human chorionic gonadotrophin in serum. *J. clin. Endocrin.* 23: 125-131.

629 Mohler, S. R., Dille, J. R. and Gibbons, H. L. (1968). The time zone

and circadian rhythms in relation to aircraft occupants taking long distance flights. *Amer. J. Publ. Health.* 58: 1404-1409.

630 MOLE, R. H. (1945). Diurnal and sampling variations in the determination of haemoglobin. *J. Physiol.* 104: 1-5.

631 MÖLLERSTRÖM, J. (1929). Om dygnsvariationer i blod-och urinsockerkurvan hos diabetiker. *Hygiea (Swed.).* 91: 379-397.

632 MÖLLERSTRÖM, J. (1933). Periodicity of carbohydrate metabolism and rhythmic functioning of the liver. Their significance in the treatment of diabetes with insulin. *Arch. intern. Med.* 52: 649-663.

633 MÖLLERSTRÖM, J. (1934). The treatment of diabetes with reference to the endogenous periodicity of the carbohydrate metabolism. *Acta. med. scand. Suppl.* 59: 145-161.

634 MÖLLERSTRÖM, J. (1935). Some new observations and principles concerning diabetes research and their practical application to diabetic therapy. *Acta med. Upsalien.* 41: 287-372.

635 MÖLLERSTRÖM, J. (1943). Das Diabetesproblem: Die rhythmischen Stoffwechselvorgänge. *Acta med. scand. Suppl.* 147: 1-476.

636 MÖLLERSTRÖM, J. (1947). Theoretical considerations and practical results in the management of diabetes. *Acta med. scand. Suppl.* 196: 12-23.

637 MÖLLERSTRÖM, J. (1954). Diurnal rhythm in severe diabetes mellitus. The significance of harmoniously timed insulin treatment. *Diabetes,* 3: 188-191.

638 MÖLLERSTRÖM, J. and SOLLBERGER, A. (1958). The 24-hour rhythm of metabolic processes in diabetes: citric acid in the urine. *Acta. med. scand.* 160: 25-46.

639 MÖLLERSTRÖM, J. and SOLLBERGER, A. (1962). Fundamental concepts underlying the metabolic periodicity in diabetes. *Ann. N.Y. Acad. Sci.* 98: 984-994.

640 MOLYNEUX, M. K. B. (1964). Use of single urine samples for the assessment of lead absorption. *Brit. J. industr. Med.* 21: 203-209.

641 MOLYNEUX, M. K. B. (1966). Observations on the excretion rate and concentration of mercury in urine. *Ann. Occup. Hyg.* 9: 95-102.

642 MORAN, L. J. and MEFFERD, R. B. (1959). Repetitive psychometric measures. *Psychol. Reports.* 5: 269-275.

643 MORLEY, A. A. (1967). The effects of cortisol on the rate of autohaemolysis of red cells incubated *in vitro. Brit. J. Haemat.* 13: 310-316.

644 MORLEY, A. A., BAKER, L. R., BEARDWELL, C. G. and BURKE, C. W. (1967). Adrenal steroids and haemolysis in paroxysmal nocturnal haemoglobinuria. *Lancet.* 2: 448-450.

645 MOSHKOV, B. S., FUKSHANSKIY, L. YA, and YUZEFOVICH, G. I. (1966). On the construction of a mathematical model of a 'biological clock'. *C.R. Akad. Sci., U.R.S.S.* 167: 440-443.

646 MOSSO, U. (1887). Récherches sur l'inversion des oscillations diurnes de la température chez l'homme normal. *Arch. ital. Biol.* 8: 177-185.

647 MOTT, P. E., MANN, F. C., MCLOUGHLIN, Q. and WARWICK, D. P. (1965). *Shift work. The Social, Psychological and Physical Consequences.* London: Cresset Press.

648 MOTTRAM, J. C. (1945). Diurnal variation in production of tumours. *J. Path. Bact.* 57: 265-267.

649 MUELLER, S. C. and BROWN, G. E. (1930). Hourly rhythms in blood pressure in persons with normal and elevated pressures. *Ann. intern. Med.* 3: 1190-1200.

650 MULLER, A. F., MANNING, E. L. and RIONDEL, A. M. (1958). The effect of potassium depletion and repletion on aldosterone excretion. *J. clin. Invest.* 37: 918.

651 MULLER, A. F., MANNING, E. L. and RIONDEL, A. M. (1958). Influence of position and activity on the secretion of aldosterone. *Lancet I*, 711-713.

652 MULLER, A. F., MANNING, E. L. and RIONDEL, A. M. (1958). Diurnal variation of aldosterone related to position and activity in normal subjects and patients with pituitary insufficiency. In *An International Symposium on Aldosterone*. Muller, A. F. and O'Connor, C. M. (eds.). London: Churchill. 111-127.

653 MULLER, A. F., RIONDEL, A. M. and MANNING, E. L. (1957). L'excrétion de l'aldostérone au cours du nychthémère. *Helv. med. acta.* 24: 463-471.

654 MULLIN, F. J. (1939). Development of the diurnal temperature and motility patterns in a baby. *Amer. J. Physiol.* 126: 589.

655 MUNCH-PETERSEN, S. (1950). The variations in serum copper in the course of 24 hours. *Scand. J. clin. Lab. Invest.* 2: 48-52.

656 MUSCIO, B. (1920). Fluctuations in mental efficiency. *Brit. J. Psychol.* 10: 327-344.

657 McCARTHY, E. F. and VAN SLYKE, D. D. (1939). Diurnal variations of haemoglobin in the blood of normal men. *J. biol. Chem.* 128: 567-572.

658 McCLURE, D. J. (1966). The diurnal variation of plasma cortisol levels in depression. *J. psychosom. Res.* 10: 189-195.

659 McEACHERN, J. M. and GILMORE, C. R. (1932). Studies in cholesterol metabolism; physiological variations in blood cholesterol. *Canad. med. Ass. J.* 26: 30-33.

660 McGIRR, P. O. M. (1966). Health considerations. In *The benefits and problems of shift working*. P.E.R.A. symposium, Sec. 5. 1-6.

661 McGIRR, P. O. M. (1967). Circadian rhythms in flight. *Trans. Soc. Occup. Med.* 18: 3-12.

662 McKEE, L. C., JOHNSON, L. E. and HEYSSEL, R. M. (1967). Circadian rhythm of plasma iron transport. *Clin. Res.* 15: 66.

663 McKENZIE, S. (1882). A case of filarial haemato-chyluria. *Trans. path. Soc., London.* 33: 394-410.

664 McKERROW, C. B., McDERMOTT, M., GILSON, J. C. and SCHILLING, R. S. F. (1958). Respiratory function during the day in cotton workers: a study in byssinosis. *Brit. J. industr. Med.* 15: 75-83.

665 NABARRO, J. D. N. (1956). The adrenal cortex and renal function. In *Modern Views on the Secretion of Urine*. Winton, F. R. (ed.). London: Churchill. 148-185.

666 NATHANIELSZ, P. W. (1967). A circadian rhythm in the disappearance of thyroxine from the blood in the calf and the thyroidectomised rat. *J. Physiol.* 204: 79-90.

667 NELSON, D. H. and SAMUELS, L. T. (1952). A method for the determination of 17-hydroxycorticosteroids in blood: 17-hydroxycorticosterone in the peripheral circulation. *J. clin. Endocrin.* 12: 519-526.

668 NELSON, D. H., SANDBERG, A. A., PALMER, J. G. and TYLER, F. H. (1952).

Blood levels of 17-hydroxycorticosteroids following the administration of adrenal steroids and their relation to levels of circulating leukocytes. *J. clin. Invest.* 31: 843-849.

669 NEUBERGER, F. and SCHMID, R. (1962). Hearing threshold and sympathetic nervous system tone, their cyclic concordance and temporary coincidence in 24-hr rhythm. *Arch. Ohr.-Nas-u. Kehlk.* 179: 237-258.

670 NEUBERGER, F. and SCHMID, R. (1962). On the hearing threshold cycle and phase relation of air conduction and bone conduction. *Arch. Ohr.-Nas.-u. Kehlk.* 179: 386-399.

671 NEWELL, F. W. and KRILL, A. E. (1965). Diurnal tonography in normal and glaucomatous eyes. *Amer. J. Ophthal.* 59: 840-853.

672 NEY, R. L., SCHIMIZU, N., NICHOLSON, W. E., ISLAND, D. P. and LIDDLE, G. W. (1963). Correlation of plasma ACTH concentration with adrenocortical response in normal human subjects, surgical patients, and patients with Cushing's Disease. *J. clin. Invest.* 42: 1669-1677.

673 NICHOLS, T., NUGENT, C. A. and TYLER, F. H. (1965). Diurnal variation in suppression of adrenal function by glucocorticoids. *J. clin. Endocrin.* 25: 343-349.

674 NICHOLS, C. T. and TYLER, F. H. (1967). Diurnal variation in adrenal cortical function. *Ann. Rev. Med.* 18: 313-324.

675 NIELSEN, L. A. (1944). On serum copper. III. Normal values. *Acta. med. scand.* 118: 87-91.

676 NOLL, V. H. (1932). A study of fatigue in 3-hour college ability tests. *J. appl. Psychol.* 16: 175-183.

677 NORN, M. (1929). Über schwankungen der Kalium-Natrium-und Chloride Ausscheidung durch die Niere im Laufe des Tages. *Skand. Arch. Physiol.* 55: 184-210.

678 NUGENT, C. A., EIK-NES, K., KENT, H. S., SAMUELS, L. T. and TYLER, F. H. (1960). A possible explanation for Cushing's syndrome associated with adrenal hyperplasia. *J. clin. Endocrin.* 20: 1259-1268.

679 NUTTALL, F. Q. and JONES, B. (1968). Creatine kinase and glutamic oxalacetic transaminase activity in serum: kinetics of change with exercise and effect of physical conditioning. *J. Lab. clin. Med.* 71: 847-854.

680 O'CONNOR, W. J. (1962). *Renal Function.* London: Arnold. Monographs of the Physiological Society, No. 10.

681 ODELL, W. D., WILBER, J. F. and UTIGER, R. D. (1967). Studies of thyrotrophin physiology by means of radioimmunoassay. *Recent Progr. Horm. Res.* 23: 47-85.

682 OGATA, K. and SASAKI, T. (1963). On the causes of diurnal body temperature rhythm in man, with reference to observations during voyage. *Jap. J. Physiol.* 13: 84-96.

683 OGLE, W. (1866). On the diurnal variations in the temperature of the human body in health. *St. George's Hosp. Rep.* 1: 221-245.

684 OLIVI, O. and GENOVA, R. (1962). Richerche sul bioritmo fisiologico dei 17, 21- Diidrossi-20-Corticosteroidi liberi plasmatici nel lattante. *Folia endocr.* 15: 421-430.

685 ORBAN, G. and CZEIZEL, E. (1967). Der Tagesrhythmus der Geburten. *Gynaecologia*, 163: 173-178.

686 ORTH, D. N., ISLAND, D. P. and LIDDLE, G. W. (1967). Experimental

alteration of the circadian rhythm in plasma cortisol (17.OHCS) concentration in man. *J. clin. Endocrin.* 27: 549-555.

687 OSBORNE, W. A. (1908). Body temperature and periodicity. *J. Physiol.* 36: 39-41P.

688 OSTERGAARD, T. (1944). The excitability of the respiratory centre during sleep and during *Evipan* anaesthesia. *Acta. physiol. scand.* 8: 1-15.

689 OURGAUD, A. G. and ETIENNE, R. (1961). Exploracion funcional del ojo glaucomatoso. *Arch. Oftal. B. Aires.* 36: 177-190.

690 PAGE, I. H. and MOINUDDIN, M. (1962). Hourly variations in serum cholesterol. *J. Atheroscler. Res.* 2: 181-185.

691 PALMER, A. (1950). The basal body temperature of women: I. Correlation between temperature and time factor. *Amer. J. Obstet. Gynec.* 59: 155-161.

692 PARIN, V. V., YR, M., VOLYNKIN, M. and VASSILYEV, P. V. (1964). *Manned Space Flight.* COSPAR symposium, Florence, Italy.

693 PARMELEE, A. H., Jr. (1961). Sleep patterns in infancy: a study of one infant from birth to eight months of age. *Acta. paediat.* 50: 160-170.

694 PARMELEE, A. H., Jr., SCHULTE, F. J., AKIYAMA, Y., WENNER, W. H., SCHULTZ, M. A. and STERN, E. (1968). Maturation of EEG activity during sleep in premature infants. *Electroenceph. clin. Neurophysiol.* 24: 319-329.

695 PARMELEE, A. H., Jr., SCHULZ, H. R. and DISBROW, M. A. (1961). Sleep patterns of the newborn. *J. Pediat.* 58: 241-250.

696 PARMELEE, A. H., Jr., WENNER, W. H., AKIYAMA, Y., SCHULTZ, M. and STERN, E. (1967). Sleep states in premature infants. *Development. Med. Child Neurol.* 9: 70-77.

697 PARMELEE, A. H., Jr., WENNER, W. H. and SCHULZ, H. R. (1964). Infant sleep patterns: from birth to 16 weeks of age. *J. Pediat.* 65: 576-582.

698 PATERSON, J. C. S., MARRACK, D. and WIGGINS, H. S. (1952). Hypoferraemia in the human subject, the importance of diurnal hypoferraemia. *Clin. Sci.* 11: 417-423.

699 PATRY, F. L. (1931). The relation of time of day, sleep and other factors to the incidence of epileptic seizures. *Amer. J. Psychiat.* 10: 789-813.

700 PEARSON, R. G. and BYARS, G. E. (1956). Development and validation of a checklist for measuring subjective fatigue. *School of Aviation Med. USAF Report No.* 56-115.

701 PEKKARINEN, A., RAURAMO, L. and THOMASSON, B. (1963). On free and conjugated 17-hydroxycorticosteroids and its diurnal rhythm in plasma during normal and toxemic pregnancy. *Acta. endocr., Copenhagen. Suppl.* 75: 1-24.

702 PELLEGRINI, L. and AGRIOLAS, L. (1950). La fosforemia nelle 24 ore. *Minerva med.* 41: 315-319.

703 PEMBREY, M. S. (1898). Animal heat. In Schafer, E. A., *Textbook of Physiology.* Edinburgh and London: Pentland. 1: 785-867.

704 PEMBREY, M. S. and NICOL, B. A. (1898). Observations upon the deep and surface temperature of the human body. *J. Physiol.* 23: 386-406.

705 PERKOFF, G. T., EIK-NES, K., NUGENT, C. A., FRED, H. L., NIMER, R. A., RUSH, L., SAMUELS, L. T. and TYLER, F. H. (1959). Studies of the diurnal variation of plasma 17-hydroxycorticosteroids in man. *J. clin. Endocrin.* 19: 432-443.

706 PETERSON, J. E., WILCOX, A. A., HALEY, M. I. and KEITH, R. A. (1960). Hourly variation in total serum cholesterol. *Circulation*, 22: 247-253.

707 PETERSON, R. E. (1957). Plasma corticosterone and hydroxycortisone levels in man. *J. clin. Endocrin.* 17: 1150-1157.

708 PETERSON, R. E. (1959). The miscible pool and turnover rate of adrenocortical steroids in man. *Recent Progr. Hormone Res.* 15: 231-274.

709 PETERSON, R. E., NOKES, G., CHEN, P. S. Jr. and BLACK, R. L. (1960). Estrogens and adrenocortical function in man. *J. clin. Endocrin.* 20: 495-514.

710 PETERSON, R. E. and WYNGAARDEN, J. B. (1956). The miscible pool and turnover rate of hydrocortisone in man. *J. clin. Invest.* 35: 552-561.

711 PETRÉN, T. and SOLLBERGER, A. (1965). Developmental rhythms. In *The Cellular Aspects of Biorhythms.* Von Mayersbach, H. (ed.). Berlin, Heidelberg, New York: Springer-Verlag.

712 PFAFF, D. (1968). Effects of temperature and time of day on time judgments. *J. exp. Psychol.* 76: 419-422.

713 PFAFF, F. (1897). Some observations in a case of human pancreatic fistula. *J. Boston Soc. med. Sci.* 2: 10-18.

714 PFAFF, F. and BALCH, A. W. (1897). An experimental investigation of some of the conditions influencing the secretion and composition of human bile. *J. exp. Med.* 2: 49-105.

715 PICKERING, G. W., SLEIGHT, P. and SMYTH, H. S. (1967). The relation of arterial pressure to sleep and arousal in man. *J. Physiol.* 191: 76-78P.

716 PIÉRON, H. (1913). *Le Problème Physiologique du Sommeil.* Paris: Masson et Cie.

717 PIÉRON, H. (1952). *The Sensations: their Functions, Processes, and Mechanisms.* New Haven: Yale University Press.

718 PINCUS, G. (1943). A diurnal rhythm in the excretion of urinary ketosteroids by young men. *J. clin. Endocrin.* 3: 195-199.

719 PINCUS, G., ROMANOFF, L. P. and CARLO, J. (1948). A diurnal rhythm in the excretion of neutral reducing lipids by man and its relation to the 17-ketosteroid rhythm. *J. clin. Endocrin.* 8: 221-226.

720 PITTENDRIGH, C. S. (1960). Circadian rhythms and the circadian organization of living systems. *Cold Spr. Harb. Symp. quant. Biol.* 25: 159-184.

721 PITTENDRIGH, C. S. and BRUCE, V. G. (1957). An oscillator model for biological clocks. In *Rhythmic and Synthetic Processes in Growth*, Rudnick (ed.), Princeton: Princeton University Press, 75-109.

722 PITTENDRIGH, C. S. and BRUCE, V. G. (1959). Daily rhythms as coupled oscillator systems and their relation to thermoperiodism and photoperiodism. In *Photoperiodism and Related Phenomena in Plants and Animals*, Washington, D.C.: Am. Assoc. Advan. Sci., Publ. 55: 475-505.

723 PITTENDRIGH, C. S., BRUCE, V. G. and KAUS, P. (1958). On the significance of transients in daily rhythms. *Proc. nat. Acad. Sci. Wash.* 44: 965-973.

724 POINTS, T. C. (1956). Twenty-four hours in a day. *Obstét. et. Gynéc.* 8: 245-248.

725 PÖPPEL, E. (1968). Desynchronisationen circadianer Rhythmen innerhalb einer isolierten Gruppe. *Pflüg. Arch. ges. Physiol.* 299: 364-370.

726 PRICE-JONES, C. (1920). The diurnal variation in the sizes of red blood cells. *J. Path. Bact.* 23: 371-383.

727 QUABBE, J. H., SCHILLING, E. and HELGE, H. (1966). Pattern of growth hormone secretion during a 24-hour fast in normal adults. *J. clin. Endocrin.* 26: 1173-1177.

728 QUAY, W. B. (1968). Individuation and lack of pineal effect in the rat's circadian locomotor rhythm. *Physiol. Behav.* 3: 109-118.

729 QUINCKE, H. (1877). Über den Einfluss des Schlafes auf die Harnabsonderung. *Arch. exp. Path. Pharmak.* 7: 115-125.

730 QUINCKE, H. (1893). Über Tag- und Nachtharn. *Arch. exp. Path. Pharmak.* 32: 211-249.

731 RABINOVITCH, I. M. (1923). Variations of the percentage of haemoglobin in man during the day. *J. Lab. clin. Med.* 9: 120-123.

732 RABINOVITCH, I. M. and STREAN, G. (1924). Hemoglobin content of the red blood cells in relation to their surface area: preliminary report. *Arch. int. Med.* 34: 124-128.

733 RABINOWITSCH, W. (1929). Untersuchungen über die alveolare Kohlensäurespannung bei natürlichem Schlaf und bei Wirkung von Schlafmitteln. *Z. ges. exper. Med.* 66: 284-290.

734 RADNÖT, M. (1962). Significance of the retina in the optico-vegetative function. *Atti VII Conf. Intern. Soc. Stud. Ritm. Biol.* Turin: Panminerva Med. 127-129.

735 RADNÖT, M. and WALLNER, E. (1964). Periodicity in the eosinophil count in the adrenal cycle. *Ann. N.Y. Acad. Sci.* 117: 244-253.

736 RADNÖT, M., WALLNER, E. and HÖENIG, M. (1960). Die Wirkung des Lichtes und des Hydergin auf die eosinophilen Leukocyten des Blutes. *Wein. Klin. Wschr.* 72: 101-105.

737 RADNÖT, M., WALLNER, E. and HÖENIG, M. (1961). Netzhautfunktion und optico—vegetatives system. *Acta. Chirung. Hung.* 2: 419-444.

738 RADNÖT, M., WALLNER, E. and TÖRÖK, E. (1956). Die Wirkung des Lichtes auf die Nebennierenrindenfunktion. *Acta. med. Hung.* 9: 231-236.

739 RANDRUP, A. and MUNKVAD, I. (1966). Changes in urine volume and diurnal rhythm caused by reserpine treatment of schizophrenic patients. *Brit. J. Psychiat.* 112: 173-176.

740 RAPOPORT, M. I. and BEISEL, W. R. (1968). Circadian periodicity of tryptophan metabolism. *J. clin. Invest.* 47: 934-939.

741 RAPOPORT, M. I., FEIGIN, R. D., BRUTON, J. and BEISEL, W. R. (1966). Circadian rhythm for tryptophan pyrrolase activity and its circulating substrate. *Science,* 153: 1642-1644.

742 RECTOR, F. C. Jr., VAN GIESEN, G., KIIL, F. and SELDIN, D. W. (1964). Influence of expansion of extracellular volume on tubular reabsorption of sodium independent of changes in glomerular filtration rate and aldosterone activity. *J. clin. Invest.* 43: 341-348.

743 REED, D. J. and KELLOGG, R. H. (1958). Changes in respiratory response to CO_2 during natural sleep at sea level and at altitude. *J. appl. Physiol.* 13: 325-330.

744 REGELSBERGER, H. (1928). Untersuchungen über die Schlafkurve des Menschen. *Z. klin. Med.* 107: 674-692.

745 REICHLING, G. H. and KLIGMAN, A. M. (1961). Alternate-day corticosteroid therapy. *Arch. Derm. Syph., N.Y.* 83: 134-137.

746 Reinberg, A. (1965). Circadiane Rhythmen und allergische Reaktion beim Menschen. *Fortschr. Med.* 11: 175-177.

747 Reinberg, A. (1966). Rythmes des fonctions cortico-surrénaliennes et systèmes circadiens. *Prob. actuels d'Endocrin. Nut.* 10: 75-89.

748 Reinberg, A. (1967). The hours of changing responsiveness or susceptibility. *Perspect. Biol. Med.* 11: 111-128.

749 Reinberg, A. and Ghata, J. (1964). *Biological Rhythms.* New York: Walker & Co.

750 Reinberg, A., Halberg, F., Ghata, J. and Siffre, M. (1966). Spectre thermique (rhythmes de la température rectale) d'une femme adult avant, pendant et après son isolement souterrain de trois mois. *C.R. Acad. Sci., Paris*, 262: 782-785.

751 Reinberg, A. and Sidi, E. (1966). Circadian changes in the inhibitory effects of an antihistaminic drug in man. *J. invest. Derm.* 46: 415-419.

752 Reinberg, A., Sidi, E. and Ghata, J. (1965). Circadian reactivity rhythms of human skin to histamine or allergen and the adrenal cycle. *J. Allergy*, 36: 273-283.

753 Reinberg, A., Zagula-Mally, Z. W., Ghata, J. and Halberg, F. (1967). Circadian rhythm in duration of salicylate excretion referred to phase of excretory rhythms and routine. *Proc. Soc. exp. Biol., N.Y.* 124: 826-832.

754 Reiss, M. (1955). Investigations into psychoendocrinology. *Int. Rec. Med.* 166: 196-203.

755 Renbourn, E. T. (1947). Variation, diurnal and over longer periods of time, in blood haemoglobin, haematocrit, plasma protein, erythrocyte sedimentation rate, and blood chloride. *J. Hyg., Camb.* 45: 455-467.

756 Renton, G. H. and Weil-Malherbe, H. (1956). Adrenaline and noradrenaline in human plasma during natural sleep. *J. Physiol.* 131: 170-175.

757 Resko, J. A. and Eik-Nes, K. B. (1966). Diurnal testosterone levels in peripheral plasma of human male subjects. *J. clin. Endocrin.* 26: 573-576.

758 Retiene, K., Schumann, G., Tripp, R. and Pfeiffer, E. F. (1966). Über das Verhalten von ACTH und Cortisol im Blut von Normalen und Kranken mit primärer und sekundärer Störung der Nebennierenrindenfunktion. II. Die Tagesrhythmik der ACTH-Sekretion im Regulationsmechanismus des Hypothalmus-hypophysen-nebennierenrinden-systems. *Klin. Wschr.* 44: 716-721.

759 Richardson, D. W., Honour, A. J., Fenton, G. W., Scott, F. H. and Pickering, G. W. (1964). Variation in arterial pressure throughout the day and night. *Clin. Sci.* 26: 445-460.

760 Richter, C. P. (1960). Biological clocks in medicine and psychiatry: Shock phase hypothesis. *Proc. nat. Acad. Sci., Wash.* 46: 1506-1530.

761 Richter, C. P. (1965). *Biological Clocks in Medicine and Psychiatry.* Springfield, Ill.: Charles C. Thomas.

762 Reiper, P. (1934). Zu viel Schlaf - zu wenig Schlaf? *Umschau.* 38: 585-586.

763 Ringer, S. and Stuart, A. P. (1877). On the temperature of the human body in health. *Proc. Roy. Soc., B.* 26: 186-210.

764 Rippmann, E. T. (1964). Die zeitliche Verteilung von 10,000 Geburten auf die 24-Stunden des Tages. *Gynaecologia*, 158: 31-34.

765 Rivlin, R. S. and Melmon, K. L. (1965). Cortisone-provoked depres-

sion of plasma thyroxine concentration: relation to enzyme induction in man. *J. clin. Invest.* 44: 1690-1698.

766 Rizzo, S. C. (1965). Variazioni nictoemerali della escrezione urinaria del Mg in soggetti normali. *Boll. Soc. ital. Biol. sper.* 41: 273-275.

767 Roberts, E. and Simonson, D. G. (1962). Free amino acids in animal tissues. In *Amino Acid Pools, Distribution, Formation and Function of Free Amino Acids.* Holden, J. T. (ed.). Amsterdam: Elsevier.

768 Roberts, H. J. (1964). Afternoon glucose tolerance testing: a key to pathogenesis, early diagnosis and prognosis of diabetogenic hyperinsulinism. *J. Am. Geriat. Soc.* 12: 423-472.

769 Roberts, W. (1860). Observations on some of the daily changes of the urine. *Edinb. med. J.* 5: 817-825.

770 Robertson, M. E., Stiefel, M. and Laidlaw, J. C. (1959). The influence of estrogen on the secretion, disposition and biologic activity of cortisol. *J. clin. Endocrin.* 19: 1381-1398.

771 Robin, E. D., Whaley, R. D., Crump, C. H. and Travis, D. M. (1958). Alveolar gas tensions, pulmonary ventilation and blood pH during physiological sleep in normal subjects. *J. clin. Invest.* 37: 981-989.

772 Romanoff, L. P., Plager, J. and Pincus, G. (1949). The determination of adrenocortical steroids in human urine. *Endocrinology,* 45: 10-20.

773 Romanoff, L. P., Rodriguez, R. M., Seelye, J. M. and Pincus, G. (1957). Determination of tetrahydrocortisol and tetrahydrocortisone in the urine of normal and schizophrenic men. *J. clin. Endocrin.* 17: 777-785.

774 Romanoff, L. P., Seelye, J., Rodriguez, R. and Pincus, G. (1957). The regular occurrence of 3α-allotetrahydrocortisol (3α, 11β, 17α-21-tetrahydroxyallopregnan-20-one) in human urine. *J. clin. Endocrin.* 17: 434-437.

775 Rosenbaum, J. D., Ferguson, B. C., Davis, R. K. and Rossmeisl, E. C. (1952). The influence of cortisone upon the diurnal rhythm of renal excretory function. *J. clin. Invest.* 31: 507-520.

776 Rosenbaum, J. D., Papper, S. and Ashley, M. M. (1955). Variations in renal excretion of sodium independent of change in adrenocortical hormone dosage in patients with Addison's disease. *J. clin. Endocrin.* 15: 1459-1474.

777 Roth, J., Glick, S. M., Yalow, R. S. and Berson, S. A. (1963). Secretion of human growth hormone: physiologic and experimental modifications. *Metabolism,* 12: 577-579.

778 Roth, J., Glick, S. M., Yalow, R. S. and Berson, S. A. (1963). Hypoglycaemia: a potent stimulus to secretion of growth hormone. *Science,* 140: 987-988.

779 Rud, F. (1947). The eosinophil count in health and in mental disease. A biometrical study. *Acta. psychiat. Kbh. Supp.* 40: 1-443.

780 Rutenfranz, J. (1955). Zur Frage einer Tagesrhythmik des elektrischen Hautwiderstandes beim Menschen. *Intern. Z. Angew. Physiol.* 16: 152-172.

781 Rutenfranz, J. (1961). The development of circadian system functions during infancy and childhood. In *Circadian Systems,* Report 39th Ross Conference on Pediatric Research, Fomon, S. J. (ed.). Columbus, Ohio: Ross Laboratories, 38-41.

782 Rutenfranz, J., Hellbrügge, T. H. and Niggeschmid, T. W. (1956).

Über die Tagesrhythmik des elektrischen Hautwiderstandes bei 11-jährigen Kindern. *Z. Kinderheilk*, 78: 144-157.

783 SABIN, F. R., CUNNINGHAM, R. S., DOAN, C. J. and KINDWALL, J. A. (1925). The normal rhythm of the white blood cells. *Bull. Johns Hopk. Hosp.* 37: 14-67.

784 SACHAR, E. J., MASON, J. W., KOLMER, H. S. and ARTISS, K. L. (1963). Psychoendocrine aspects of acute schizophrenic reactions. *Psychosom. Med.* 25: 510-537.

785 SANDBERG, A. A., NELSON, D. H., GLENN, E. M., TYLER, F. H. and SAMUELS, L. T. (1953). 17-hydroxycorticosteroids and 17-ketosteroids in urine of human subjects: clinical application of a method employing β-glucuronidase hydrolysis. *J. clin. Endocrin.* 13: 1445-1464.

786 SANDBERG, A. A., NELSON, D. H., PALMER, J. G., SAMUELS, L. T. and TYLER, F. H. (1953). The effects of epinephrine on the metabolism of 17-hydroxycorticosteroids in the human. *J. clin. Endocrin.* 13: 629-647.

787 SANDWEISS, D. J., FRIEDMAN, M. H. F., SUGARMAN, M. H. and PODOLSKY, H. M. (1946). Nocturnal gastric secretion. II. Studies on normal subjects and patients with duodenal ulcer. *Gastroenterology*, 7: 38-54.

788 SANDWEISS, D. J., SUGARMAN, M. H., PODOLSKY, H. M. and FRIEDMAN, M. H. F. (1946). Nocturnal gastric secretion in duodenal ulcer. Studies on normal subjects and patients with their bearing on ulcer management. *J. Amer. med. Ass.* 130: 258-265.

789 SASAKI, T. (1964). Effect of rapid transposition around the earth on diurnal variation in body temperature. *Proc. Soc. exp. Biol., N.Y.* 115: 1129-1131.

790 SCHAEFER, K. E., CLEGG, B. R., CAREY, C. R., DOUGHERTY, J. H. and WEYBREW, B. B. (1967). Effect of isolation in a constant environment on periodicity of physiological functions and performance levels. *Aerospace Med.* 38: 1002-1018.

791 SCHAEFER, K. E. and DOUGHERTY, J. H. Jr. (1966). Variability of respiratory functions based on circadian cycles. *SMRL Report No.* 486. *U.S.N. Submarine Medical Center, Groton, Conn.*

792 SCHÄFER, K. H. and BOENICKE, I. (1949). Die neurovegetative Lenkung des Eisenstoffwechsels. *Arch. exp. Path. Pharmak.* 207: 666-687.

793 SCHALLY, A. V., ANDERSEN, R. N., LIPSCOMB, H. S., LONG, J. M. and GUILLEMIN, R. (1960). Evidence for the existence of two corticotrophin-releasing factors, α and β. *Nature, Lond.* 188: 1192-1193.

794 SCHATZ, D. L. and VOLPÉ, R. (1959). Lack of diurnal variation in the level of serum protein-bound iodine. *J. clin. Endocrin.* 19: 1495-1497.

795 SCHEVING, L. E. (1957). Mitotic activity in human epidermis. *Anat. Rec.* 127: 363.

796 SCHEVING, L. E. (1959). Mitotic activity in the human epidermis. *Anat. Rec.* 135: 7-14.

797 SCHINDL, I. (1952). Das antidiuretische Prinzip (ADP) bei hydropischen Zustanden. *Helv. Med. Acta.* 19: 238-257.

798 SCHLAPHOFF, D., JOHNSTON, F. A. and BOROUGHS, E. D. (1950). Serum iron levels of adolescent girls and the diurnal variation of serum iron and hemoglobin. *Arch. Biochem.* 28: 165-173.

799 SCHLEGEL, B. (1967). Chronophysiologie und chronopathologie. *Verh. dtsch. Gesell. Inn. Med.* 73: 886-1117.

800 SCHMITT, O. H. (1960). Biophysical and mathematical models of circadian rhythms. *Cold Spr. Harb. Symp. quant. Biol.* 25: 207-210.

801 SCHMITT, O. H. (1961). Oscillatory systems as models of periodicity. In *Circadian Systems*. Report 39th Ross Conference on Pediatric Research, Fomon, S. J. (ed.). Columbus, Ohio: Ross Laboratories. 27-28.

802 SCHMITT, O. H. (1962). Adaptive analog models for biological rhythms. *Ann. N.Y. Acad. Sci.* 98: 846-850.

803 SCHREUDER, O. B. (1966). Medical aspects of aircraft pilot fatigue with special reference to the commercial jet pilot. *Aerospace Med.* 37 Suppl.: 1-30.

804 SCHWEIG, (1843). Untersuch. über periodische Vorgänge etc. Quoted in Speck, (1882). *Arch. exp. Path. Pharmakol.* 15: 81-145.

805 SEAMAN, G. V., ENGEL, R., SWANK, R. L. and HISSEN, W. (1965). Circadian periodicity in some physicochemical parameters of circulating blood. *Nature, Lond.* 207: 833-835.

806 SEGRE, E. J. and KLAIBER, E. L. (1966). Therapeutic utilization of the diurnal variation in pituitary-adrenocortical activity. *Calif. West. Med.* 104: 363-365.

807 SHARP, G. W. G. (1960). Reversal of diurnal rhythms of water and electrolyte excretion in man. *J. Endocrin.* 21: 97-106.

808 SHARP, G. W. G. (1960). Reversal of diurnal leucocyte variations in man. *J. Endocrin.* 21: 107-114.

809 SHARP, G. W. G. (1960). The effect of light on diurnal leucocyte variations. *J. Endocrin.* 21: 213-218.

810 SHARP, G. W. G. (1960). The effect of light on the morning increase in urine flow. *J. Endocrin.* 21: 219-223.

811 SHARP, G. W. G. (1961). Reversal of diurnal temperature rhythms in man. *Nature, Lond.* 190: 146-148.

812 SHARP, G. W. G. (1962). Simultaneous reversal of diurnal rhythms of urine and electrolyte excretion, leucocytes and sleep/wakefulness. *Atti. VII Conf. Intern. Soc. Stud. Ritm. Biol.* Turin: Panminerva Med. 133-138.

813 SHARP, G. W. G., SLORACH, S. A. and VIPOND, H. J. (1961). Diurnal rhythms of keto-and ketogenic steroid excretion and the adaptation to changes of the activity-sleep routine. *J. Endocrin.* 22: 377-385.

814 SHAW, A. F. B. (1927). The diurnal tides of the leucocytes of man. *J. Path.* 30: 1-19.

815 SHERWOOD, J. J. (1965). A relation between arousal and performance. *Amer. J. Psychol.* 78: 461-465.

816 SHETTLES, L. B. (1960). Hourly variation in onset of labor and rupture of membranes. *Amer. J. Obstet. Gynec.* 79: 177-179.

817 SHOLITON, L. J., WERK, E. E. and MARNELL, R. T. (1961). Diurnal variation of adrenocortical function in non-endocrine disease states. *Metabolism*, 10: 632-646.

818 SIFFRE, M. (1965). *Beyond Time* (transl. by Briffault, H.). London: Chatto and Windus.

819 SIFFRE, M., REINBERG, A., HALBERG, F., GHATA, J., PERDRIEL, G. and SLIND, R. (1966). L'isolement souterrain prolongé—étude de deux sujets

adultes sains avant, pendant et après cet isolement. *Pr. méd.* 74: 915-919.

820 SIMMONS, D. H., ASSALI, N. A. and AREDON, M. (1960). Relative influence of respiratory and metabolic acid-base changes on renal acid excretion. *Amer. J. Physiol.* 198: 237-243.

821 SIMPSON, A. S. (1952). Are more babies born at night? *Brit. med. J.* 2: 831.

822 SIMPSON, G. E. (1924). Diurnal variations in the rate of urine excretion for two hour intervals—some associated factors. *J. biol. Chem.* 59: 107-122.

823 SIMPSON, G. E. (1926). The effect of sleep on urinary chloride and pH. *J. biol. Chem.* 67: 505-516.

824 SIMPSON, G. E. (1929). Changes in the composition of urine brought about by sleep and other factors. *J. biol. Chem.* 84: 393-411.

825 SIMPSON, H. W. Personal communication.

826 SIMPSON, H. W. and LOBBAN, M. C. (1967). Effect of a 21-hour day on the human circadian excretory rhythms of 17-hydroxycorticosteroids and electrolytes. *Aerospace Med.* 38: 1205-1213.

827 SIROTA, H., BALDWIN, D. S. and VILLARREAL, H. (1950). Diurnal variation of renal function in man. *J. clin. Invest.* 29: 187-192.

828 SISAKYAN, N. M., and YAZDOVSKIY, V. I. (1964). First group flight into outer space, 11-15 Aug, 1962. *U.S. Dept. of Comm. JBRS*, 25: 272.

829 SMITH, J. L., STEMPFEL, R. S., CAMPBELL, H. S., HUDNELL, A. B. and RICHMAN, D. W. (1962). Diurnal variation of plasma 17-hydroxycorticosteroids and intraocular pressures in glaucoma. *Amer. J. Ophthal.* 54: 411-418.

830 SOLLBERGER, A. (1965). *Biological Rhythm Research.* New York: Elsevier Publishing Co.

831 SOLLBERGER, A. (1967). Biological measurements in time, with particular reference to synchronization mechanisms. *Ann. N.Y. Acad. Sci.* 138: 561-599.

832 SOLLBERGER, A., APPLE, H. P., GREENWAY, R. M., KING, P. H., LINDAN, O. and RESWICK, J. B. (1965). Automation in biological rhythm research with special reference to studies on homo. In *The Cellular Aspects of Biorhythms*, Von Mayersbach, H. (ed.). Berlin, Heidelberg and New York: Springer-Verlag.

833 SOLLBERGER, A. and PETRÉN, T. (1961). *Rept. V. Intern. Conf. Soc. Biol. Rhythm.* Stockholm: Aco-print.

834 SOUTHERN, A. L., GORDON, G. G., TOCHIMOTO, S., PINZON, G., LANE, D. R. and STYPULKOWSKI, W. (1967). Mean plasma concentration, metabolic clearance and basal plasma production rates of testosterone in normal young men and women using a constant infusion procedure: effect of time of day and plasma concentration on the metabolic clearance rate of testosterone. *J. clin. Endocrin.* 27: 686-694.

835 SOUTHERN, A. L., TOCHIMOTO, S., CARMODY, N. C. and ISURUGI, K. (1965). Plasma production rates of testosterone in normal adult men and women and in patients with the syndrome of feminizing testes. *J. clin. Endocrin.* 25: 1441-1450.

836 SOYKA, L. F. and SAXENA, K. M. (1965). Alternate-day steroid therapy for nephrotic children. *J. Amer. med. Assoc.* 192: 225-230.

837 SPANGLER, R. A. and SNELL, F. M. (1961). Sustained oscillations in a catalytic chemical system. *Nature, Lond.* 191: 457-458.
838 SPECK, B. (1968). Diurnal variation of serum iron and the latent iron-binding in normal adults. *Helv. med. acta.* 34: 231-238.
839 STAMM, D. (1967). Tagesschwankungen der normalbereiche diagnostisch wichtiger Blutbestandteile. *Verh. dtsch. ges. Inn. Med.* 73: 982-989.
840 STANBURY, S. W. (1958). Some aspects of disordered renal tubular function. *Advanc. intern. Med.* 9: 231-282.
841 STANBURY, S. W. and THOMSON, A. E. (1951). Diurnal variations in electrolyte excretion. *Clin. Sci.* 10: 267-293.
842 STARKWEATHER, W. H., SPENCER, H. H., SCHWARZ, E. L. and SCHOCH, H. K. (1966). The electrophoretic separation of lactate dehydrogenase isoenzymes and their evaluation in clinical medicine. *J. Lab. clin. Med.* 67: 329-343.
843 STEINMETZ, P. R. and EISENGER, R. P. (1966). Influence of posture and diurnal rhythm on the renal excretion of acid: observations in normal and adrenalectomised subjects. *Metabolism,* 15: 76-87.
844 STEPHENS, G. J. and HALBERG, F. (1965). Human time estimation. *Nursing Research,* 14: 310-317.
845 STERKY, G. C. G., PERSSON, B. E. H. and LARSSON, Y. A. A. (1966). Dietary fats, the diurnal blood lipids and ketones in juvenile diabetes. *Diabetologia,* 2: 14-19.
846 STRAUB, H. (1915). Alveolargasanalysen. I. Über Schwankungen in der Tätigkeit des Atemzentrums, speziell im Schlafe. *Dtsch. arch. klin. Med.* 117: 397-418.
847 STRUGHOLD, H. (1952). Physiological day-night cycle in global flights. *J. Aviat. Med.* 23: 464-473.
848 STRUGHOLD, H. (1962). Day-night cycling in atmospheric flight, space flight, and on other celestial bodies. *Ann. N.Y. Acad. Sci.* 98: 1109-1115.
849 STRUGHOLD, H. (1965). The physiological clock in aeronautics and astronautics. *Ann. N.Y. Acad. Sci.* 134: 413-422.
850 STRUMWASSER, F. (1965). The demonstration and manipulation of a circadian rhythm in a single neuron. In *Circadian Clocks,* Aschoff, J. (ed.), Amsterdam: North-Holland. 442-462.
851 STRUMWASSER, F. (1967). Types of information stored in single neurons. In *Invertebrate Nervous Systems,* Wiersma (ed.), Chicago and London: U. Chic. Press.
852 SWANSON, J. N., BAUER, W. and ROPES, M. (1952). The evaluation of eosinophil counts. *Lancet,* 1: 129-132.
853 SZCZEPANSKA, E., PREIBISZ, J., DRZEWIECKI, K. and KOZLOWSKI, S. (1968). Studies of the circadian rhythm of variations of the blood antidiuretic hormone in humans. *Pol. med. J.* 7: 517-523.
854 TÄHTI, E. (1956). Studies of the effect of x-radiation on 24-hour variations in the mitotic activity in human malignant tumours. *Acta. path. microbiol. scand. Suppl.* 117: 1-61.
855 TAKEBE, K., SETAISHI, C., HIRAMA, M., YAMAMOTO, M. and HORIUCHI, Y. (1966). Effects of a bacterial pyrogen on the pituitary-adrenal axis at various times in the 24 hours. *J. clin. Endocrin.* 26: 437-442.
856 TALIAFERRO, I., COBEY, F. and LEONE, L. (1956). Effect of diethylstil-

bestrol on plasma 17-hydroxycorticosteroid levels in humans. *Proc. Soc. exp. Biol., N.Y.* 92: 742-744.

857 Tatai, K. and Ogawa, A. (1951). A study of diurnal variation in circulating eosinophils, especially with reference to sleep in healthy individuals. *Jap. J. Physiol.* 1: 328-331.

858 Taylor, D. M., Parker, R. P., Field, E. O., Greatorex, C. A. (1968). An interpretation of the results of measurements of the uptake of ^{32}P in human tumours. *Brit. J. Radiol.* 41: 432-439.

859 Taylor, P. J. (1967). Shift and day work. A comparison of sickness absence, lateness, and other absence behaviour at an oil refinery from 1962 to 1965. *Brit. J. industr. Med.* 24: 93-102.

860 Teleky, L. (1943). Problems of nightwork: influences on health and efficiency. *Indust. Med. Surg.* 12: 758-779.

861 Teoh, E. S. (1967). Chorionic gonadotrophin in the serum and urine of Asian women in normal pregnancy. *J. Obstet. Gynaec. Brit. Comm.* 74: 74-79.

862 Teoh, E. S. (1967). Immunological diagnosis of hydatidiform mole. *J. Obstet. Gynaec. Brit. Comm.* 74: 80-84.

863 Tharp, G. D. and Folk, G. E. Jr. (1965). Rhythmic changes in rate of the mammalian heart and heart cells during prolonged isolation. *Comp. Biochem. Physiol.* 14: 255-273.

864 Thayer, R. E. (1967). Measurement of activation through self-report. *Psychol. Rep.* 20: 663-678.

865 Thiel, R. (1925). Die physiologischen und experimentell erzeugten Schwankungen des intraoculären Drückes im gesunden und glaukomatöser Augen. *Arch. Augenheilk.* 96: 331-354.

866 Thiss-Evenson, E. (1958). Shift work and health. *Indust. Med.* 27: 493-497.

867 Thomassen, T. L. (1947). The venous tension in eyes suffering from simple glaucoma. *Acta. Ophthal.* 25: 221-241.

868 Thomassen, T. L. (1947). On aqueous veins. *Acta Ophthal.* 25: 369-376.

869 Thor, D. H. (1962). Diurnal variability in time estimation. *Percept. mot. Skills.* 15: 451-454.

870 Tiedt, N. (1963). Die 24-stunden-rhythmik der kinetik des lichtreflexes der menschlichen pupille. *Pflüg. Arch. ges. Physiol.* 277: 458-472.

871 Tingley, J. O., Morris, A. W. and Hill, S. R. (1958). Studies on the diurnal variation and response to emotional stress of the thyroid gland. *Clin. Res. Proc.* 1: 134.

872 Tiovanen, P. Harri, J. and Kalliomaki, J. L. (1963). Daily changes in serum glutamic-oxalacetic acid transaminase, with reference to the clinical aspects. *Cardiologia, Basle.* 42: 391-394.

873 Toole, J. F. (1968). Nocturnal strokes and arterial hypotension. *Ann. intern. Med.* 68: 1132-1133.

874 Toulouse, E. and Piéron, H. (1907). Le méchanisme de l'inversion chez l'homme du rhythme nycthéméral de la température. *J. Physiol. Path. gén.* 9: 425-440.

875 Tränkle, W. (1961). *Rhythmologische Abläufe als psychophysischer Ausdruck.* Rep. 5th Conf. Soc. Biol. Rhythm. Stockholm: ACO-print, 167-172.

876 Tune, G. S. (1969). Sleep and wakefulness in 509 normal human adults. *Brit. J. med. Psychol.* 42: 75-80.
877 Tune, G. S. (1969). Sleep and wakefulness in a group of shiftworkers. *Brit. J. industr. med.* 26: 54-58.
878 Turner, K. B. and Steiner, A. (1939). A long-term study of the variation of serum cholesterol in man. *J. clin. Invest.* 18: 45-49.
879 Tyler, F. H., Migeon, A., Florentin, H. H. and Samuels, L. T. (1954). The diurnal variation of 17-hydroxycorticosteroid levels in plasma. *J. clin. Endocrin.* 14: 774.
880 Utterback, R. A. and Ludwig, G. D. (1949). A comparative study of schedules for standing watches aboard submarines based on body temperature cycles. *Naval Med. Res. Inst., Bethesda, Md. Rept.* No. 1. *Proj. NM* 004 003.
881 Vagnucci, A. I., Hesser, M. E., Kozak, G. P., Pauk, G. L., Lauler, D.P. and Thorn, G. W. (1965). Circadian cycle of urinary cortisol in healthy subjects and in Cushing's syndrome. *J. clin. Endocrin.* 25: 1331-1339.
882 Vagnucci, A. I. and Wesson, L. G. (1964). Diurnal cycle of renal hemodynamics and excretion of chloride and potassium in hypertensive subjects. *J. clin. Invest.* 43: 522-531.
883 Vahlquist, B. C. (1941). Das Serumeisen. Eine pädiatrischklinische und experimentelle Studie. *Acta paediat., Stockh.* 28 suppl. 5: 1-68.
884 Val, D. F. C. (1942). The night secretion of free hydrochloric acid in the stomach. *Ill. med. J.* 81: 149-152.
885 Van der Pol, B. (1940). Biological rhythms considered as relaxation oscillations. *Acta med. Scand. Supp.* 108: 76-88.
886 van Loon, J. H. (1963). Diurnal body temperature curves in shift workers. *Ergonomics*, 6: 267-273.
887 Venning, E. H., Dyrenfurth, I. and Giroud, C. J. P. (1956). Aldosterone excretion in healthy persons. *J. clin. Endocrin.* 16: 1326-1332.
888 Vernon, H. M. (1940). *The Health and Efficiency of Munition Workers.* London: Oxford University Press.
889 Visscher, M. B. and Halberg, F. (1955). Daily rhythms in numbers of circulating eosinophils and some related phenomena. *Ann. N.Y. Acad. Sci.* 59: 834-849.
890 Vogel, J. (1854). Klinische Untersuchungen über den Stoffwechsel bei gesunden und kranken Menschen überhaupt, und den durch den Urin in besondere. *Arch. des Ver. f. gem. Arb.* 1. 96. In *Speck* (1882) *Arch. f. exper. Path. u. Pharmakol.* 15: 81-145.
891 Voigt, E. D., Engel, P. and Klein, H. (1967). Daily fluctuations of the performance-pulse index. *Germ. med. Mon.* 12: 394-395.
892 Voigt, E. D., Engel, P. and Klein, H. (1968). Über den Tagesgang der körperlichen Leistungsfähigkeit. *Int. Z. Angew Physiol.* 25: 1-12.
893 Völker, H. (1927). Über die tagesperiodischen Schwankungen einiger Lebensvorgänge des Menschen. *Pflüg. Arch. ges. Physiol.* 215: 43-77.
894 von Domarus, A. (1931). Die Bedeutung der Kammerzählung der Eosinophilen für die Klinik. *Dtsch. Arch. Klin. Med.* 171: 333-358.
895 von Eiff, A. W., Böckh, E. M., Göpfert, H., Pfleiderer, F. and Steffen, T. (1953). Die Bedeutung des Zeitbewusstseins für die 24

Stunden-Rhythmen des erwachsenen Menschen. *Z. ges. exp. Med.* 120: 295-307.

896 von Euler, U. S. (1956). *Noradrenaline.* Springfield, Illinois: C. C. Thomas.

897 von Euler, U. S., Hellner-Bjorkman, S. and Orwen, I. (1955). Diurnal variations in the excretion of free and conjugated noradrenaline and adrenaline in urine from healthy subjects. *Acta. physiol. scand.* 33: Supp. 118: 10-16.

898 von Mayersbach, H. (1965). *The Cellular Aspects of Biorhythms.* Berlin, Heidelberg, New York: Springer-Verlag.

899 Voutilainen, A. (1953). Über die 24-Stunden-Rhythmik der Mitosenfrequenz in malignen Tumoren. *Acta. path. microbiol. scand. Supp.* 99: 1-104.

900 Wade, L. (1955). 'Human relations' approach to sickness, absenteeism and other employee problems. *Arch. industr. Health.* 12: 592-608.

901 Waldenström, J. (1946). The incidence of 'iron deficiency' (sideropenia) in some rural and urban populations. *Acta. med. scand. Suppl.* 170: 252-279.

902 Walfish, P. G., Britton, A., Melville, P. H. and Ezrin, C. (1961). A diurnal pattern in the rate of disappearance of I^{131} labeled l-thyroxine from the serum. *J. clin. Endocrin.* 21: 582-586.

903 Walford, J., Lammers, B., Schilling, R. S. F., van den Hoven van Genderen, D. and van der Veen, Y. G. (1966). Diurnal variation in ventilatory capacity. An epidemiological study of cotton and other factory workers employed on shift work. *Brit. J. industr. Med.* 23: 142-148.

904 Walker, J. (1966). Frequent alternation of shifts on continuous work. *Occup. Psychol.* 40: 215-225.

905 Wallace, E. Z., Silverberg, H. I. and Carter, A. C. (1957). Effect of ethinylestradiol on plasma 17-hydroxycorticosteroids, ACTH responsiveness and hydrocortisone clearance in man. *Proc. Soc. exp. Biol.* N.Y. 95: 805-808.

906 Walsh, R. J., Arnold, B. J., Lancaster, H. O., Coote, M. A. and Cotter, H. (1953). A study of haemoglobin values in New South Wales. *Spec. Rep. Ser. No.* 5. *Nat. Hlth. and Res. Coun., Australia.*

907 Ward, H. C. (1904). The hourly variations in the quantity of hemoglobin and in the number of the corpuscles in human blood. *Amer. J. Physiol.* 11: 394-403.

908 Ward, W. D. (1964). Diurnal variability of auditory threshold. *Acta. oto-laryng., Stockh.* 58: 139-142.

909 Warter, J. S. (1866). Remarks on the use of the thermometer in disease. *St. Bart's Hosp. Rep.* 2: 64-79.

910 Webb, W. B. and Agnew, H. (1964). Reaction time and serial response efficiency on arousal from sleep. *Percept. Mot. Skills.* 18: 783-784.

911 Wedderburn, A. A. I. Personal communication.

912 Weil-Malherbe, H. (1955). The concentration of adrenaline in human plasma and its relation to mental activity. *J. Ment. Sci.* 101: 733-755.

913 Weitzman, E. D., Schaumburg, H. and Fishbein, W. (1966). Plasma 17-hydroxycorticosteroid levels during sleep in man. *J. clin. Endocrin.* 26: 121-127.

914 Welsh, J. H. (1938). Diurnal rhythms. *Quart. Rev. Biol.* 13: 123-139.

915 WESSON, L. G. (1964). Electrolyte excretion in relation to diurnal cycles of renal function. *Medicine*, 43: 547-592.

916 WESSON, L. G. and LAULER, D. P. (1961). Diurnal cycle of glomerular filtration rate and sodium and chloride excretion during responses to altered salt and water balance in man. *J. clin. Invest.* 40: 1967-1977.

917 WEVER, R. (1960). Possibilities of phase-control, demonstrated by an electronic model. *Cold Spr. Harb. Symp. quant. Biol.* 25: 197-206.

918 WEVER, R. (1962). Zum Mechanismus der biologischen 24-Stunden-Periodik. *Kybernetik*, 1: 139-154.

919 WEVER, R. (1964). Zum Mechanismus der biologischen 24-Stunden-Periodik III. Mitteilung: Anwendung der Modell-Gleichung. *Kybernetik*, 2: 127-144.

920 WEVER, R. (1968). Gesetzmässigkeiten der circadianen Periodik des Menschen, geprüft an der Wirkung eines schwachen elektrischen Wechselfeldes. *Pflüg. Arch. ges. Physiol.* 302: 97-122.

921 WEVER, R. (1968). Mathematical models of circadian rhythms and their applicability to men. In *Cycles Biologiques et Psychiatrie*, de Ajuriaguerra, J. (ed.). Symposium Bel-Air III. Geneva: Georg. 61-72.

922 WEVER, R. (1969). Autonome circadiane Periodik des Menschen unter dem Einfluss verschiedener Beleuchtungs-Bedingungen. *Pflüg. Arch. ges. Physiol.* 306: 71-91.

923 WEYSSE, A. W. and LUTZ, B. R. (1915). Diurnal variations in blood pressure. *Amer. J. Physiol.* 37: 330-347.

924 WHITTAKER, E. T. and ROBINSON, G. (1926). *The Calculus of Observations*, 2nd Ed. London: Blackie & Son.

925 WILKERSON, H. L. C. and KRALL, L. P. (1957). Diabetes in a New England town. *J. Am. Med. Ass.* 135: 209-216.

926 WILKINSON, R. T., FOX, R. H., GOLDSMITH, R., HAMPTON, I. F. G. and LEWIS, H. E. (1964). Psychological and physiological responses to raised body temperature. *J. appl. Physiol.* 19: 287-291.

927 WILSON, R. H. L., NEWMAN, E. J. and NEWMAN, H. W. (1956). Diurnal variation in rate of alcohol metabolism. *J. appl. Physiol.* 8: 556-558.

928 WINKELSTEIN, A. (1935). One-hundred-sixty-nine studies in gastric secretion during the night. *Amer. J. dig. Dis.* 1: 778-782.

929 WOLF, W. (1962). Rhythmic functions in the living system. *Ann. N.Y. Acad. Sci.* 98: 753-1326.

930 WOLFE, L. K., GORDON, R. D., ISLAND, D. P. and LIDDLE, G. W. (1966). An analysis of factors determining the circadian pattern of aldosterone excretion. *J. clin. Endocrin.* 26: 1261-1266.

931 WOLFF, H. P. and TORBICA, M. (1963). Determination of plasma-aldosterone. *Lancet*, 1: 1346-1348.

932 WOODHEAD, G. S. and VARRIER-JONES, P. C. (1916). Investigations on clinical thermometry. Continuous and quasi-continuous temperature records in man and animals in health and disease. *Lancet*, 1: 173-180.

933 WOOLLEY-HART, A., TWENTYMAN, P., CORFIELD, J., JOSLIN, C., MORRISON, R. and FOWLER, J. F. (1968). Changes in ^{32}P counting-rate in human and animal tumours. *Brit. J. Radiol.* 41: 440-447.

934 WRONG, O. and DAVIS, H. E. F. (1959). The excretion of acid in renal disease. *Quart. J. Med.* 28: 259-313.

935 WURTMAN, R. J. (1967). Ambiguities in the use of the term circadian. *Science*, 156: 104.

936 WURTMAN, R. J. Personal communication.

937 WURTMAN, R. J., CASPER, A., POHORECKY, L. A. and BARTTER, F. C. (1968). Impaired secretion of epinephrine in response to insulin among hypophysectomized dogs. *Proc. nat. Acad. Sci., Wash.* 61: 522-528.

938 WURTMAN, R. J., CHOU, C. and ROSE, C. M. (1967). Daily rhythm in tyrosine concentration in human plasma: persistence on low-protein diets. *Science*, 158: 660-662.

939 WURTMAN, R. J., ROSE, C. M., CHOU, C., and LARIN, F. F. (1968). Daily rhythms and the concentration of various amino acids in human plasma. *New Engl. J. Med.* 279: 171-175.

940 WURTMAN, R. J., ROSE, C. M., ROSE, L., WILLIAMS, G. and LAWLER, D. (1968). Diurnal amino-acid rhythms in endocrine diseases. *Clin. Res.* 16: 355.

941 WYATT, S. and MARRIOTT, R. (1953). Night work and shift changes. *Brit. J. industr. Med.* 10: 164-172.

942 WYATT, S. and WESTON, H. C. (1920). A performance test under industrial conditions. *Brit. J. Psychol.* 10: 293-309.

943 YERUSHALMY, J. (1938). Hour of birth and stillbirth and neonatal mortality rates. *Child Dev.* 9: 373-378.

944 ZSÓTÉR, T. and SEBÖK, S. (1955). Daily variation of antidiuretic substance in the blood serum. *Acta. med. scand.* 152: 47-52.

945 ZÜLCH, K. J. and HOSSMANN, V. (1967). 24-hour rhythm of human blood pressure. *Ger. med. Mon.* 12: 513-518.

Index